Calculus Laboratories
Using DERIVE

L. Carl Leinbach
Gettysburg College

Wadsworth Publishing Company
Belmont, California
A Division of Wadsworth, Inc.

Mathematics Editor: Anne Scanlan-Rohrer

Editorial Assistant: Leslie With

Managing Designer: Andrew Ogus

Print Buyer: Barbara Britton

Cover Designer: Rob Hugel/xxx Design

Cover Photographer: Kerrick James

Printed in the United States of America   49

1 2 3 4 5 6 7 8 9 10—95 94 93 92 91

ISBN 0-534-15480-8

Contents

Preface

What is it that we are doing when we teach calculus as a laboratory course? To many students and instructors, it merely means that "a computer is being used." While this description describes some of the activities in the course, it does not describe all of them, nor does it convey the spirit of the course. A laboratory experience involves the following activities:

1. Observation of Phenomena

2. Identification of Some Underlying Principle

3. Experimentation with the Objects of the Investigation

4. Explanation of the Behavior Observed

5. Analysis of the Phenomena

The appropriate goal for the instructor is to involve the student in activities 4 and 5 through a structured use of activities 1, 2, and 3. Automatic computation has provided instructors with a means of efficiently conducting these activities. A Computer Algebra System such as DERIVE provides the ideal software tool. It is possible to conduct an experiment, modify the input, and repeat the experiment with very little bother. The student can make hypotheses and test them. Mathematical analysis must follow the laboratory activity. The lab, if properly conducted, enhances the analysis. It does not replace it. Mathematical novices have a place to "hang their hats."

The laboratory exercises in this book are of three types:

1. Those that anticipate results to be covered in class (Labs 2, 3, and 9, for example)

2. Those that expand upon material already presented in class (Labs 4, 6, and 8)

3. Interesting applications of the calculus (labs 5, 7, 10, and 11).

Each lab is designed so that the student may concentrate on the process or concept, while DERIVE does the heavy, manipulative work. A student who pays attention to the lab will be learning the essence of calculus and its power as an intellectual tool, not a collection of meaningless manipulations that are soon to be forgotten.

Although the symbolic capabilities of DERIVE are important tools for conducting an investigation, the primary asset is DERIVE's graphics capabilities. It is the picture of the function's behavior or the relationship of the graph of a derived function to the original that will make the strongest impression on the student. For this reason, several illustrations are presented in each chapter, and the student is encouraged to draw and analyze many graphs. The very first lab deals with setting up DERIVE to do graphing and a brief introduction to using the cross hair to locate points on the graph of a function. In Lab 8 we show the student some of the limitations of using graphics as the sole tool for analysis. A few correct words may be more valuable than a picture in some cases.

DERIVE as a Computer Algebra System is very close to a "walk up and use" system. We do not spend pages describing the details of using the system. In most cases a few sentences about the appropriate key strokes are enough to provide the student with the necessary information on how to use the system. Once again, the interface does not interfere with the learning process.

The overall goal of the laboratory experience is to lead students to some deep and interesting mathematics and applications of mathematics. Notice that the verb is to *lead*, not to *tell*. With DERIVE students have the power to investigate phenomena and formulate concepts. The job of the text and the instructor is to ensure that the student discovers the correct concept. Once a student has formulated a concept, it is his or hers. These labs are designed to aid in that formulation process and to make the students owners of some beautiful mathematics and powerful problem solving tools.

This manual was produced using EXP. The ease of use of this extremely versatile system made the design and layout of the material a joy. The ease with which graphical displays can be imported into the text and the vast library of mathematical symbols allowed me to say what I wanted to say and not compromise because of the limitations of the word processing system.

I greatly appreciate the cooperation of David Stoutmeyer and his staff (particularly his wife, Karen) at *The Soft Warehouse*. I have been a beta-tester for DERIVE since the dark ages of version 1.?. Dave and his staff have been extremely helpful and responsive to my "Is it possible to have DERIVE do . . . " questions. They supplied me with a beta version of DERIVE 2.0 at about the time this manual was being completed. I have indicated in a very few places some extensions that this new version of DERIVE makes possible. I am sure that when you as a student or instructor become familiar with this version, you will see many more extensions that the programming ability of version 2 makes possible. I think that the excitement has just begun!

I also thank David Stoutmeyer for reading a draft of the manuscript for technical accuracy on the use of DERIVE. In addition to saving me from some technical blunders, he made many useful suggestions that found their way into the text.

I would be completely remiss if I did not acknowledge the support that I have been given in the completion of this project. First, I appreciate the support of the staff at Wadsworth Publishing Company who have encouraged and at certain times prodded me to meet my deadlines. I particularly enjoyed working with former mathematics editor Barbara Holland, and her successor on this project,

Anne Scanlan-Rohrer. Pat Brewer was an outstanding aide as she assisted me in preparing the final manuscript. Their support and assistance were invaluable. My secretary (and friend) Lowanda Bowers, who would drop everything to help me meet a deadline, deserves a crown more than thanks. My colleagues were very helpful in evaluating these materials. They also allowed me to teach them to their students. Their feedback and comments are always taken very seriously. Particular thanks to Professor Marvin Brubaker of Messiah College, my "partner in crime." I sometimes have trouble noting where Marv's ideas end and mine start. Fortunately, Marv is generous and lets me take credit that should be his. I only hope that our association profits him as much as it does me. Finally, to my wife, Pat, who has had to live with me throughout the process of planning, preparing, writing, and rethinking my approach, I dedicate this manual.

Calculus Laboratories Using DERIVE

Laboratory #0
Setting Up DERIVE and Using Its Graphic Capabilities

Introduction

Throughout these laboratories you will be using the DERIVE® Computer Algebra System (CAS). What this means is that you will be using a program that can solve equations, do algebraic manipulations and many of the technical manipulations of calculus. Unless you instruct the system to do otherwise, it will do arithmetic in exactly the same way you and I do arithmetic. If you ask the system to solve the equation $x^2 - 2 = 0$, the response will be: $x = \sqrt{2}$ and $x = -\sqrt{2}$, not the numerical approximations $x = 1.414$ and $x = -1.414$ that we have become used to as the output for computer programs. More importantly, DERIVE "knows" how to perform algebraic operations involving symbols. It can factor equations, add fractions, and expand products. Furthermore, it can generally do this faster than you and I can do the same operations.

In addition to its symbolic capabilities, DERIVE can also make numerical approximations, draw the graphs of functions, and display data. Thus, if you really need to know that $\sqrt{2}$ can be approximated by 1.41421356, DERIVE will give you the information. However, if you use $\sqrt{2}$ in any future calculations, they will not suffer from roundoff error. The reason is that DERIVE continues to use $\sqrt{2}$ and not its numerical approximation in all calculations.

The symbolic and numerical capabilities of DERIVE are impressive, but, as has been said so many times, "A picture is worth a thousand words." In our case, the picture can give clearer meaning to the results of DERIVE's symbolic and numerical work. By examining the graphical results, we can gain insights into more general results and appreciate their reasonableness.

The following sections will teach you how to gain access to the DERIVE system, set up the screen, and draw graphs. It is assumed that your computer is set to access DERIVE, i.e., that the DERIVE disk is in your default drive or that your instructor has written commands that enable the program to be run when you type DERIVE.

Starting DERIVE and Setting Up

Begin the program by typing Derive and pressing the Return or Enter key on the right hand side of the keyboard. After a very few seconds, you should see a display that is similar to the one shown in Figure $0-1$.

```
COMMAND: Author Build Calculus Declare Expand Factor Help Jump soLve Manage
         Options Plot Quit Remove Simplify Transfer moVe Window approX
Enter option
                            Free:100%            Derive Algebra
```

Figure 0-1: The Initial DERIVE Screen

This display identifies the program, the version of the program you are running, the copyright, and intellectual ownership of the program. The help option is somewhat limited and refers you to pages in the manual for the program. Of real interest are the two lines that lie below the thick line crossing the screen. This is a list of the major DERIVE commands. These commands can be executed by either typing the capital letter that appears in the command name or using the space bar to move the highlight over the command name (the Backspace key reverses the highlight movement) and then pressing the Return key. Let's illustrate this process by creating a split screen.

In computing terminology this is called *creating two windows.* Press the "w" key for Window. At this point a new list of commands appears on the command line. Type "s" for Split and the "v" for Vertical. We accept the default that splits the screen in the middle by pressing the Enter key one more time. A split screen appears with the 1 in the upper left-hand corner highlighted. In order to make the screen on the right the **graphics window**, press the $\boxed{\text{F1}}$ key. This results in the 2 being highlighted, and now all commands will refer to the window on the right. The results of the key strokes described in this paragraph are illustrated in Figure $0-2$.

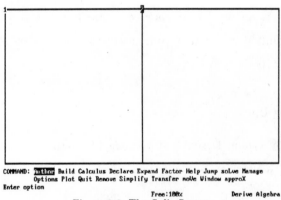

```
COMMAND: Author Build Calculus Declare Expand Factor Help Jump soLve Manage
         Options Plot Quit Remove Simplify Transfer moVe Window approX
Enter option
                            Free:100%            Derive Algebra
```

Figure 0-2: The Split Screen

To finish creating our graphics window, we type "w" followed by "d" for Designate. Remember, we are working with the window on the right. We now type "2" for 2-Dimensional Graphics. Finish the process by typing "d" for Display, "o" for Options, and "g" for Graphics followed by Enter. The result should appear much like Figure $0-3$. If your display is different, see your instructor or lab manager to correctly set up the proper DERIVE options.

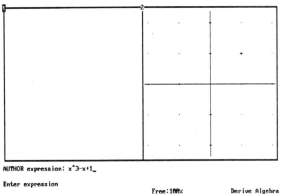

AUTHOR expression: x^3-x+1_

Enter expression

Free:100%   Derive Algebra

Figure 0-3: One Algebra Window and One Graphics Window

We are now ready to draw a graph. Press the F1 key again to highlight the 1, indicating that you are ready to work in the window on the left, the **Algebra window**.

Using DERIVE's Graphics

We begin by entering an expression. For our illustration, we will graph $x^3 - x + 1$. Press the "a" key for Author. This results in a different version of the command line. We are instructed to Author an expression. We do this by entering the expression in one dimensional form using the standard computing syntax. This situation is illustrated in Figure $0-3$ above.

After we enter the expression and press the Enter key, DERIVE displays our expression in the window on the left. We do not have to interpret the computing syntax form of the expression; it is displayed in the standard 2-Dimensional textbook format. Note also that a number is assigned to the expression. This means that if at a later time during the session we need to refer to this expression, we can do so by typing #1. A screen similar to the one you should be seeing is given in Figure $0-4$.

The graph can be drawn by once again pressing the F1 key to move to the graphics window and pressing "p" for Plot. This command always plots the expression in the algebra window that is currently highlighted. The result is shown in Figure $0-5$.

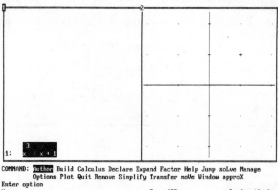

COMMAND: author Build Calculus Declare Expand Factor Help Jump soLve Manage
        Options Plot Quit Remove Simplify Transfer moVe Window approX
Enter option
User   Free:100%   Derive Algebra

Figure 0-4: DERIVE Displays Expressions in 2-D Format

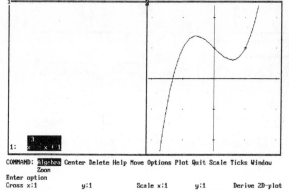

COMMAND: Algebra Center Delete Help Move Options Plot Quit Scale Ticks Window
        Zoom
Enter option
Cross x:1   y:1   Scale x:1   y:1   Derive 2D-plot

Figure 0-5: The Same Display After Pressing F1 and P for Plot

In later labs you will learn to change the scale of the graph, zoom in and out, and view different portions of the graph. One option that we will mention now is the Delete option. It allows us to remove the graph we have just drawn. Experiment with this option by pressing "d" and then "a" for All. The graph is erased. Now press "p" for Plot, and the graph is redrawn.

Finding the Coordinates of Points on the Graph

To this point nothing has been said about the cross hair that appears in the graphics window. The cross hair can be moved around on the screen using the arrow keys, the Page Up and Page Down keys, and the Ctrl-right arrow or Ctrl-left arrow combination of keys. On the right-hand side of the bottom line of the screen the x- and y-coordinates corresponding to the position of the cross hair are displayed. Graphics windows are always initialized in DERIVE with the cross hair at coordinate (1,1). Note Figure $0-4$ above. It is possible to move the cross-hair to any position on the screen. The arrow keys move the cursor in small increments. If we hold down the Ctrl key and press the right or left arrow key, the cursor moves in the appropriate horizontal direction in larger increments. The Page Up and Page Down keys move the cursor in the appropriate vertical direction in larger increments. Use these keys to find the approximate coordinates of the peak that lies to the left of the y-axis. The results of one attempt are given in Figure $0-6$. The approximation in this case is (-0.5972, 1.3928). You may have a different and even better approximation.

The reason we can only obtain an approximation is that the cross hair must move some finite increment that is dictated by the graphic display. Thus, it is not possible to move the cursor to the exact position of the peak. In fact the graph that is displayed may not pass through the exact coordinates of the peak of the idealized graph of the function. Why? Even the screen with the best possible resolution can only display a finite number of points, or pixels, as they are called in computing terminology. We know that the idealized graph of the function contains a continuum of points. Thus, the graph that is displayed on the screen is only an approximation of the true graph of the function.

Screen displays are limited, but they often give us a good idea of the behavior of the function. As we learn more calculus, we will be able to locate the exact position of the peak, but for now we must rely on our screen display and eyesight.

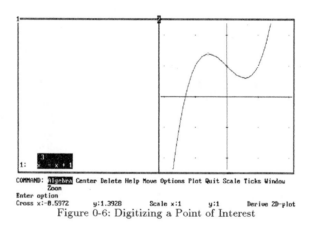

COMMAND: Algebra Center Delete Help Move Options Plot Quit Scale Ticks Window
Zoom
Enter option
Cross x:-0.5972   y:1.3928   Scale x:1   y:1   Derive 2D-plot
Figure 0-6: Digitizing a Point of Interest

DERIVE 2.0

Piecewise Defined Functions

Consider a function defined in the following way

$$f(x) = \begin{cases} x^2 - 1 & \text{if } x \leq -1 \\ \\ 2x + 2 & \text{if } x > -1 \end{cases}$$

Although the expression for f(x) is not given by a single algebraic expression, it passes our requirement for a function. In particular, for each value of the independent variable x, there is only one value for the variable, f(x). It is obvious that to the left and right of $x = -1$, the function is continuous and very "well behaved," but what happens when $x = -1$? For example, is f(x) continuous at this point, or does it have a break in the graph? If there is no break in the graph, how smoothly do the two pieces of the function join? Does the graph change direction at this point?

All of these questions can be answered analytically, but DERIVE version 2.0 and above gives us the capability of graphing functions having a piecewise definition by using a structure called an "if" statement. It's use is quite natural and almost like speaking English. For example, Authoring and Plotting the expression

$$\text{if}(\, x <= -1,\, x\hat{\ }2 - 1,\, 2x \, + 2\,)$$

yields the graph.

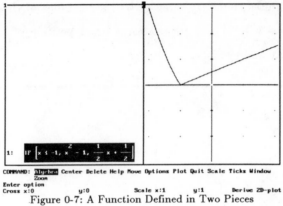

Figure 0-7: A Function Defined in Two Pieces

The syntax of the "if" statement is

$$\text{if}(<condition>,\ <expression>,\ <alternative\ expression>\,)$$

The translation of the statement is straightforward:

> *Given a condition, such as $x \leq -1$ in the previous example, then perform the action indicated by the first expression. If the condition is false, $x > -1$ in the previous example, perform the alternative action indicated by the alternative expression. If, for some reason, there is no desired alternative action, place a question mark, ?, in the alternative action field.*

In the exercises given on page 7 you will have an opportunity to experiment with piecewise defined functions. Note that it is possible to place another "if" statement as either the action or the alternative action. Thus it is possible to define functions in more than just two pieces.

Investigation of Piecewise Defined Functions

1. Using the DERIVE "if" statement, plot the following functions.

a. $f(x) = \begin{cases} x^2 & \text{if } x < 0 \\ \sqrt{x} & \text{if } x \geq 0 \end{cases}$

b. $f(x) = \begin{cases} 2 & \text{if } x < -1 \\ -2x & \text{if } -1 \leq x \leq 1 \\ -2 & \text{if } x > 1 \end{cases}$

c. $f(x) = \begin{cases} x^2 & \text{if } x < 0 \\ x + 1 & \text{if } x \geq 0 \end{cases}$

d. $f(x) = \begin{cases} -1 & \text{if } x \leq 0 \\ x + 2 & \text{if } 0 < x < 1 \\ 1 & \text{if } x \geq 1 \end{cases}$

Work Sheet

Laboratory #0

1. Using the screen display of $x^3 - x + 1$, find approximations to the following:

    a. The coordinates of the valley of the graph that lies to the right of the y-axis

    b. The root of $x^3 - x + 1$ (the x-coordinate of the point where the graph crosses the x-axis)

2. Check the accuracy of your approximation in 1b by returning to the algebra window (press the $\boxed{\text{F1}}$ key) and press "L" for soLve. You will see three answers. Use the arrow key to highlight the answer that does not have an imaginary part (the first of the three answers that are given) and press "x" for approXimate. Compare this answer to the one you obtained in 1b above.

3. Clear the Graphics window by "Deleting All" and then plot the graph of $x^2 - 5$. Find the roots of this expression. How do your approximations compare with $\pm\sqrt{5}$?

Laboratory #1
The Effects of Transformations on the Graph of a Function

Introduction

In this lab we will use the graphing and screen-control technique from Lab 0. If you have not done this lab, read the lab now to learn how to enter DERIVE, set up a graphics window, and draw the graphs of expressions. In addition to the commands covered in Lab 0, you will also be instructed in the use of the Zoom option and the Scale option as you conduct your investigations.

The purpose of this lab is to investigate the effect of certain transformations on the graph of a function. The expression defining a function determines the shape of the graph of the function. For example, a linear function in one variable has a graph that is a straight line, a quadratic function has a graph that is a parabola, and a cubic has certain distinctive shapes. We say that a graph that oscillates in a periodic fashion is *sinusoidal*. The character of the expression tells us something about the shape of the graph.

We will be considering the effects of adding a nonzero constant to a function, replacing the variable by the variable plus a constant, multiplying the function by a constant, and replacing the variable by a constant times the variable. For example, suppose our function is $f(x) = x^2$ and the constant is 2. What is the relation of the graphs of $x^2 + 2$, $(x + 2)^2$, $2x^2$, and $(2x)^2$ to the graph of the original function, $f(x) = x^2$? Figure $1-1$ shows the graphs of x^2, $x^2 + 2$, and $(x + 2)^2$. Figure $1-2$ shows the graphs of x^2, $2x^2$, and $(2x)^2$. The graphs of all of these functions are different, but they are all related in some way to the original function, $f(x)$. The question is: In what way is the graph of the transformed function related to the original graph? By the end of this lab you should be able to supply the correct answer.

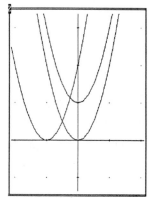

Figure 1-1: x^2 and Two Translates

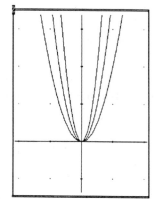

Figure 1-2: x^2 and Two Dilations

Preparations for the Investigation

We will use five functions as paradigms for our investigation. They include examples of polynomial, algebraic, rational, and trigonometric functions. During the exposition, one of the functions will be used as an illustration, and you will be asked to perform similar operations on the other functions. If the example set is rich enough, you should be able to make some general observations about the effect of applying the transformations to a function.

The functions we will use for our investigation are

1. $f(x) = x^3 - x + 1$
2. $f(x) = x^4 - x^2$
3. $f(x) = \sqrt{x^2 - 1}$
4. $f(x) = \dfrac{x}{x^2 + 2}$
5. $f(x) = \sin(x)$

Although these functions represent a broad spectrum of function types, they do not encompass all possible types of functions, nor can we make any general conclusions based solely on the behavior of these particular examples. What we observe in the lab can only serve as an aid to our intuition. We can make conjectures about certain behaviors, but we must then be able to prove our conjectures before we draw any general conclusions.

Enter the above expressions to obtain the following DERIVE display:

$$
\begin{aligned}
&1: \quad x^3 - x + 1 \\[6pt]
&2: \quad x^4 - x^2 \\[6pt]
&3: \quad \sqrt{(x^2 - 1)} \\[6pt]
&4: \quad \frac{x}{x^2 + 2} \\[6pt]
&5: \quad \text{SIN}\,(x)
\end{aligned}
$$

Figure 1-3: DERIVE Display of the Five Expressions

The square root operator is entered by pressing the Alt-q combination. Thus, the third expression is authored on the command line as: Alt-q (x^2 − 1). The parentheses are required to indicate the scope of the square root operator.

Now that you have entered these expressions, the functions can be referred to as #1, #2, #3, #4, and #5, respectively. We will begin by considering the effect of adding a constant to a function.

The Relation of the Graph of f(x) and the Graph of f(x) + a

Our example function will be: $f(x) = x^3 - x + 1$. The right-hand side was entered as expression #1. Using the up arrow key, highlight this expression. Assuming that you have a split screen with an algebra and a graphics window, press $\boxed{F1}$ and "p" to plot the graph of f(x). This is the same graph that you investigated in Lab 0. Returning to the algebra window, author the following expression: #1 + 2. When you press return, you will see the expression for f(x) + 2 is given as expression #6. This is a convenient bit of shorthand that is built into DERIVE. If you want DERIVE to perform an algebraic simplification of this expression, press "s" for Simplify and then press the Enter key.

Return to the graphics window and plot this expression. Unfortunately, much of the graph does not appear in the window. This situation can be remedied by using the Zoom command. Press "z" followed by "y" , "o", and the Enter key. This sequence instructs DERIVE to zoom out on the y-axis, i.e., to show more y values. The results of all of these activities are shown in Figure 1 − 4.

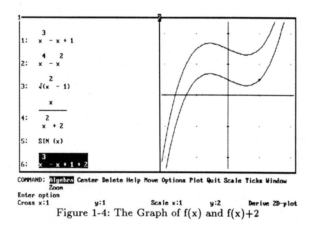

Figure 1-4: The Graph of f(x) and f(x)+2

What is the relationship of the graph of f(x)+2 to the graph of f(x)? Does it lie above or below the graph of f(x)? Look at corresponding points that lie on the same vertical line. Is there a fixed relationship between the y-values? By how much does the y-value of a point on one graph differ from the y-value of a point directly above it on the other graph? What are the answers to these questions if some other value is added to expression #1? What if a negative constant is added to the expression?

Investigate the results of adding nonzero constants to the other functions that you are considering. Clear the graphics window before investigating each function. This is done by pressing "d" for Delete and "a" for All. After completing these investigations, can you answer the questions in the previous paragraph for a general function, f(x), and a constant, a?

The Relation of the Graph of f(x + a) and the Graph of f(x)

To create the expression for f(x + a), highlight expression #1. Plot the graph and return to the algebra window. Press "m" for Manage and "s" for Substitute. The command line will show

SUBSTITUTE expression: #1

You press the Enter key. The command line then shows

SUBSTITUTE VALUE: x

You type x+2 and press the Enter key. The expression for f(x) should appear with x replaced by x+2. If this is not the case, remove the current expression, check to see that your keyboard is in overwrite mode (look near the lower right hand corner of the DERIVE screen to determine whether or not you are in overwrite mode). If Insert appears, press the Insert key, and try again. Now plot this expression.

Once again, many of the interesting points of the graph do not appear on the screen. To remedy this, zoom out on the x-axis. When the graphs have reappeared, answer the following questions. What is the relationship of the graph of f(x+2) to the graph of f(x)? Does it lie to the left or the right of the graph of f(x)? What about points that lie directly beside each other on the same horizontal line? By how much do their x-coordinates differ? What are the answers if we look at f(x+a) for some other values of a? What if a is negative?

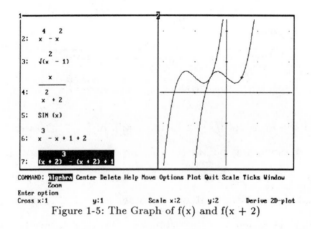

Figure 1-5: The Graph of f(x) and f(x + 2)

After investigating the graphs of f(x + a) and f(x) for some of the other functions given in our list, what conjecture can be made about the relationship of the graph of f(x + a) to the graph of f(x)?

The Graphs of a·f(x) and f(a·x)

For this part of the lab we will illustrate with the function f(x) = sin(x). We begin by highlighting #5 and plotting the expression. Now author 2*#5 and plot the result.

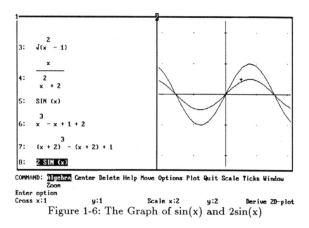

Figure 1-6: The Graph of sin(x) and 2sin(x)

Notice that the graph of 2·f(x) still has the same basic shape, but it appears to be "stretched." What is the y-value of a peak for f(x) and the y-value for a valley? What is the difference between these two values? What are the answers to the same questions for 2·f(x)? For the sine function, one-half of this difference is called the <u>amplitude</u>. What is the amplitude of a·sin(x)?

Sin(x) is a bounded function. If the function is not bounded, it does not make sense to talk about amplitude, but what about the distances between successive peaks and valleys on the graphs of f(x) and a·f(x)? Is there any relationship?

Finally, clear the graphics window and in the algebra window highlight and plot expression #5 again. Using the Manage and Substitute options, replace x by 2x in the expression (or just author sin(2x)). Plot this expression. Note that the two graphs are the same height, but one is more compressed than the other. Which one is compressed and by how much is it compressed? Compare the corresponding horizontal distances between the points where the graphs cross the x-axis. What is the relationship between the corresponding distances? What about the horizontal distances between successive peaks and valleys? What is the relationship between these distances on the graph of f(x) and f(2x)? (See Figure 1 − 7 below.)

Since sin(x) is periodic, the answers to the question of the relationship of the graphs is more obvious than the general case. However, a comparison of the graphs of f(x) and f(a·x) should convince you that this transformation has an effect similar to that you observed for sin(x). For example, what happens to the distances between successive peaks and valleys? How much horizontal distance is required to cover the same vertical distance? Answers to these questions can be suggested by examining the graphs of the example functions for particular values of a.

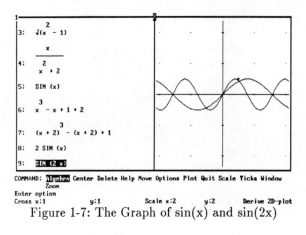

Figure 1-7: The Graph of sin(x) and sin(2x)

The Laboratory Report

Write an essay that describes the effects of the transformations you explored in this lab. In particular, describe the relationship of the graph of f(x) to the graphs of f(x)+a, f(x+a), a·f(x), and f(a·x). What is your evidence for your conclusions? You may include illustrations with your written text. Use as your starting point the answers to the questions posed throughout the writeup of this lab. Write as if you are explaining your observations to a friend. Concentrate on describing what you observed and not the mechanical operations you performed to have DERIVE generate a display. Your report should be written in clear, coherent language that can be understood by an individual who has not read the lab writeup or conducted the investigations.

Independent Investigation

The graph of the function, $f(x) = \sin(x) + \cos(x)$ is shown in Figure $1 - 8$.

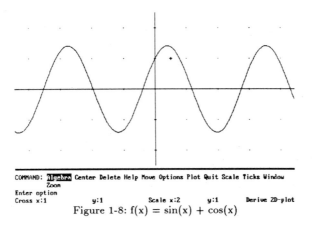

COMMAND: Algebra Center Delete Help Move Options Plot Quit Scale Ticks Window
         Zoom
Enter option
Cross x:1   y:1   Scale x:2   y:1   Derive 2D-plot

Figure 1-8: $f(x) = \sin(x) + \cos(x)$

Notice that this graph appears to have the same shape as the graph of the sine function. Its amplitude is greater than 1, and it is shifted to the left. Find constants, a and b, such that $f(x) = a\cdot\sin(x + b)$. Use the point-digitizing capabilities of DERIVE to make your estimates for a and b and then plot the graphs of $\sin(x) + \cos(x)$ and $a\cdot\sin(x + b)$ using your estimates for a and b. Does the graph of the second function overlay the graph of $f(x)$?

Work Sheet

Laboratory #1

1. Let $f(x) = \dfrac{x}{x^2 + 2}$. On the axes below draw the graphs of $f(x)$, $f(x) + 2$, $f(x) - 1$, $f(x+2)$, and $f(x - 1)$.

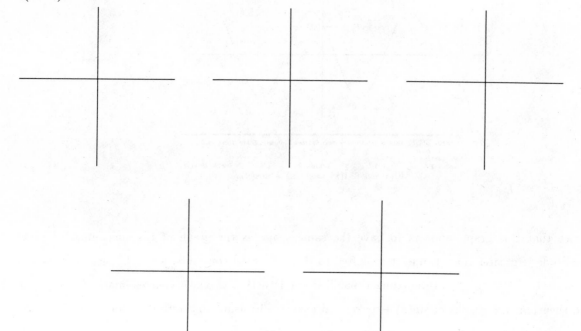

  a. How do transformations of the form $f(x) + a$ appear to affect

   (1) The shape of the graph?

   (2) The position of the graph?

  b. How do transformations of the form $f(x + a)$ appear to affect

   (1) The shape of the graph?

   (2) The position of the graph?

2. Let $f(x) = \sin(x)$. On the axes below draw the graphs of $f(x)$, $2 \cdot f(x)$, and $f(2 \cdot x)$.

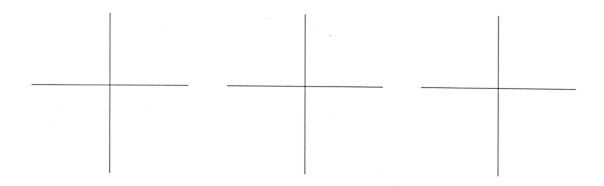

    a. How do transformations of the form $a \cdot \sin(x)$ affect the shape of the sine graph?

    b. How do transformations of the form $\sin(a \cdot x)$ affect the shape of the sine graph?

3. Repeat the above exercise, except this time use the function $f(x) = x^3 - x + 1$.

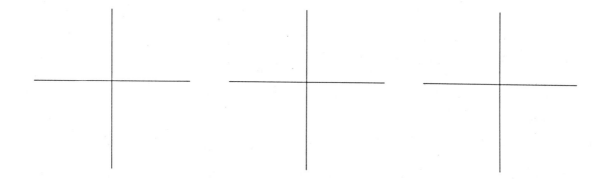

    a. What do you think is the general effect of the transformation $a \cdot f(x)$?

    b. What do you think is the general effect of the transformation $f(a \cdot x)$?

Laboratory #2

The Relationship Between a Function and Its Derivative

Introduction

In calculus courses we define the tangent line to the graph of a function, f(x), at the point $(x_0, f(x_0))$ to be the limiting position of the secant lines drawn through $(x_0, f(x_0))$. In Figure $2-1$ we show the graph of a function, $f(x)=2^x$, the tangent to this graph at the point $(1,2)$, and the secant line through the points $(1,2)$ and $(\frac{3}{2}, 2\sqrt{2})$. Note that if we were to choose a point on the graph closer to $(1,2)$ the secant line will be even more tightly aligned with the tangent line.

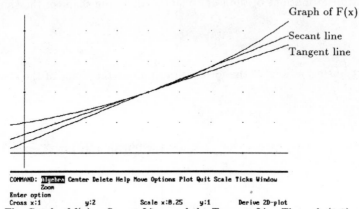

Graph of F(x)

Secant line

Tangent line

```
COMMAND: Algebra Center Delete Help Move Options Plot Quit Scale Ticks Window
         Zoom
Enter option
Cross x:1          y:2          Scale x:8.25   y:1        Derive 2D-plot
```

Figure 2-1: The Graph of f(x), a Secant Line, and the Tangent Line Through (1,2)

Since the tangent and secant lines both pass through the point $(1,2)$, it is the slopes that define the differences in the lines. Thus, saying that the tangent line is the limiting position of the secant lines means that the slope of the tangent line is the limit of the slopes of the secant lines, i.e., if we choose some "small" value for $h \neq 0$, and let $m_h(x_0)$ be the slope of the secant line through $(x_0, f(x_0))$ and $(x_0+h, f(x_0+h))$, then m, the slope of the tangent line at $(x_0, f(x_0))$, is approximated by m_h. Thus,

$$m \approx m_h(x_0) = \frac{f(x_0+h) - f(x_0)}{h}$$

or,

$$m = \lim_{h \to 0} m_h(x_0) = \lim_{h \to 0} \frac{f(x_0+h) - f(x_0)}{h}$$

The definition of the slope is intimately tied to the definition of the function f itself. To examine this relationship, we fix h as .001 and allow x to vary. This defines a function, $m_{.001}(x)$, whose value at each x approximates the slope, m(x), of the tangent line to the graph at x.

A Function "Derived" from f(x)

After entering DERIVE and defining an algebra and graphics window, set up the "apparatus" for our experiment. First we alert DERIVE that the letter "f" will be used to designate a function during the course of our investigation. Press "d" for Declare and then "f" for Function. When the prompt requests a function name, type "f". In response to the request for a function value, press the Enter key. DERIVE then asks for the function variables. Since f is a function of one variable, type "x" in response to the first request and press the Enter key in response to the second request. The expression, F(x):= , appears on the screen.

Having reserved the letter "f" for our function, define m(x,h) to be the slope of the secant line through (x,f(x)) and (x + h,f(x + h)). Press "d" for Declare and respond "m" for the function name. At the function value prompt type

$$(f(x + h) - f(x))/h$$

The screen display will appear as:

$$M(x,h) := \frac{F(x + h) - F(x)}{h}$$

We are now ready to begin our investigation. The prototype function will be $f(x) = x^3 - x + 1$.

Use the Declare mechanism to define the function, f(x). At the request for the function value, type: x^3 − x + 1 and press the Enter key. The function definition will immediately appear on the screen. Plot this function in the graphics window (set the y-scale to 2). Now, author the expression M(x,.001). Press "e" for Expand and plot the result. The display will be similar to that of Figure 2 − 2.

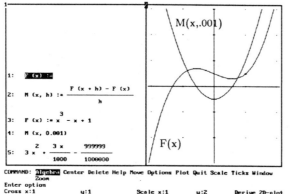

Figure 2-2: The Graph of $f(x) = x^3 - x + 1$ and the Function M(x,.001)

What is the relation between the two graphs? A natural starting point is to look at the distinguished points on the graph of f. What is happening to the derived function when the graph of f(x) has a peak or a valley? What is the behavior of f(x) when the derived function is negative? What is its behavior when the derived function is positive? Why does this make sense geometrically? Summarize your answers to the above questions by completing the following table.

Derived function	Behavior of f(x)
Zero	
Positive	
Negative	

If these relationships hold in general, then it is possible to predict the shape of f(x) by studying the derived function of f(x).

As a final step in the investigation, find the expression for m(x). Return to the algebra window and type "m(x,h)". Now press "c" for Calculus and "l" for Limit. Press the Enter key when a request is made for the expression. The response to the "LIMIT variable:" prompt is "h", followed by pressing the Enter key at the "LIMIT: POINT" request since we want the limit as h approaches 0. Press "s" for Simplify, and the result will appear after a short time. Note that the result is a quadratic, as suggested by the fact that the graph of M(x,.001) was a parabola.

The Laboratory Report

As a result of the above investigation, make a conjecture about the relationship between f(x) and its derived function. Test your conjecture by performing a similar investigation for the following functions.

1. $f(x) = \dfrac{x}{x^2 + 1}$

2. $f(x) = x^2 - 5x + 2$
3. $f(x) = \sin(x)$
4. $f(x) = \cos(x)$

Do these investigations support your conjecture? Explain why you believe that this conjecture makes sense, i.e., how does the behavior of the function dictate the nature of the slope of the tangent line and vice versa? Explain how, if you are given information about the derivative, you can use this information to determine the shape of the function.

DERIVE 2.0

Using the "If" Statement to Visualize the Action of the Function

The DERIVE "if" statement can be used to help us visualize the action of the function based on the sign of the derivative. This action can be visualized by plotting only those parts of f(x) for which the derivative satisfies a particular condition.

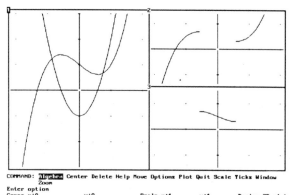

COMMAND: **Algebra** Center Delete Help Move Options Plot Quit Scale Ticks Window
Zoom
Enter option
Cross x:0   y:0   Scale x:1   y:1   Derive 2D-plot

Figure 2-3: The Graph of Figure 2-2 Dissected According to the Sign of its Derivative

In this figure the graph in Window 2 was drawn by plotting the expression

$$if(\; dif(f(x),x) > 0, \; f(x), \; ?)$$

Translating this statement as a piecewise defined function, we have

$$g(x) \; = \; \begin{cases} f(x) & \text{if } \dfrac{df(x)}{dx} > 0 \\[2em] \text{undefined} & \text{otherwise} \end{cases}$$

The graph in Window 3 was drawn by plotting the expression

$$if(\; dif(f(x),x) < 0, f(x), ?\,)$$

This display is more dramatic if you have a color screen and can superimpose the two plots using a different color for each plot. In either case, you have a visual representation of the material needed to fill in the charts on your laboratory work sheet.

Work Sheet

Laboratory #2

For each of the functions given below, find $\lim\limits_{h \to 0} M(x, h)$, the derived function for f. Draw the graphs of f(x) and its derived function on the axes given below the function definitions and fill in the tables to the right of the graphs.

$$f(x) := \frac{x}{x^2 + 1} \qquad\qquad \lim\limits_{h \to 0} M(x, h) \; =$$

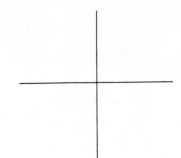

Derived function	Behavior of f(x)
Zero	
Positive	
Negative	

$$f(x) := x^2 - 5x + 2 \qquad\qquad \lim\limits_{h \to 0} M(x, h) \; =$$

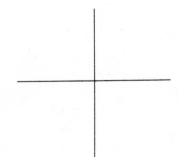

Derived function	Behavior of f(x)
Zero	
Positive	
Negative	

$f(x) := \sin(x)$

$\lim\limits_{h \to 0} M(x, h) \ =$

Derived function	Behavior of f(x)
Zero	
Positive	
Negative	

$f(x) := \cos(x)$

$\lim\limits_{h \to 0} M(x, h) \ =$

Derived function	Behavior of f(x)
Zero	
Positive	
Negative	

Laboratory #3

Exploring Rules for Differentiation

Introduction

Given functions f and g we can, under suitable conditions on the range and domain of f and g, define new functions according to the following rules.

$$(f+g)(x) \ = \ f(x) + g(x)$$

$$(f-g)(x) \ = \ f(x) - g(x)$$

$$(f \cdot g)(x) \ = \ f(x) \cdot g(x)$$

$$\left(\frac{f}{g}\right)(x) \ = \ \frac{f(x)}{g(x)}$$

$$(f \circ g)(x) = f(g(x))$$

Using the definition of the derivative,

$$f'(x) \ = \ \lim_{h \to 0} \frac{f(x+h) - f(x)}{h}$$

and the properties of the limit it is an easy matter to show that

$$(f+g)'(x) \ = \ f'(x) + g'(x)$$

One proof is given here.

$$(f+g)'(x) \ = \ \lim_{h \to 0} \frac{(f+g)(x+h) - (f+g)(x)}{h}$$

$$= \ \lim_{h \to 0} \frac{f(x+h) + g(x+h) - f(x) - g(x)}{h}$$

$$= \ \lim_{h \to 0} \frac{f(x+h) - f(x)}{h} + \lim_{h \to 0} \frac{g(x+h) - g(x)}{h}$$

$$= \ f'(x) + g'(x)$$

A similar proof shows that $(f-g)'(x) \ = \ f'(x) - g'(x)$.

The proofs of the above properties of the derivative are very straightforward, but what about the derivatives of the product, f·g, quotient, $\frac{f}{g}$, and composition, $f \circ g$, of functions? The proofs and results in these cases are less obvious than those for the sum and difference. The object of the investigation in this lab will be to use DERIVE to provide evidence for making a conjecture about the derivatives of these combinations of f and g.

During the course of this investigation, you will use the Differentiate option on the Calculus menu. To become familiar with this option, author x^n, press "c" for Calculus, and "d" for Differentiate. Press the Enter key in response to all other requests and simplify the result.

page 24

Setup for the Investigation

In order to conduct this investigation, we will need to develop a library of functions and their derivatives. You have already entered x^n into this library. Repeat the procedure to complete the following table. Some of these functions are new to you, but our emphasis is on the pattern of the derivative. Later in the course we will concern ourselves with verifying the results that DERIVE supplies in response to your commands.

$f(x)$	$\frac{d}{dx}(f(x))$
x^n	nx^{n-1}
$\sqrt{x-1}$	
$\sin(x)$	
$\mathrm{atan}(x)$	
$\mathrm{asin}(x)$	
$\ln(x)$	

Keep this chart in a handy place as you conduct the following investigations. You will need to refer to these results.

Products of Functions

We will begin this investigation by looking at the result of multiplying $\sin(x)$ by powers of x and taking the derivative. Let's assume that DERIVE has assigned #3 to the expression "$\sin(x)$". We will use that number in our demonstration.

The process we will follow is to differentiate $x \cdot \sin(x)$, $x^2 \sin(x)$, . . . , $x^5 \sin(x)$; study the results to find a pattern; make a conjecture about $x^n \sin(x)$; and test the conjecture against the DERIVE result. You will then test the conjecture against the DERIVE result of differentiating $f(x) \cdot g(x)$ for several choices of f and g from the table given above.

Begin this process by pressing "c" for Calculus and "d" for Differentiate. Respond to the requests in the following ways.

DIFFERENTIATE expression: "x∗ #3" (your responses are shown in quotes for our
demonstration. Recall #3 is "$\sin(x)$")

DIFFERENTIATION variable: "x"   (this is the default; you can press Enter)

DIFFERENTIATION order: "1"   (also a default; press Ctrl-Enter)

Pressing the Ctrl-Enter combination for the last response causes DERIVE to automatically simplify the result and saves you an additional step. Repeat the process for $x^2\sin(x)$, . . . , $x^5\sin(x)$.

The DERIVE results for differentiating for the above procedures are shown in Figure $3-1$. What pattern do you see? What do you conjecture for the result of differentiating $x^n\sin(x)$? Test your conjecture. After obtaining the simplified result press "e" for Expand.

1:   $x \cos(x) + \sin(x)$

2:   $x^2 \cos(x) + 2 x \sin(x)$

3:   $x^3 \cos(x) + 3 x^2 \sin(x)$

4:   $x^4 \cos(x) + 4 x^3 \sin(x)$

5:   $x^5 \cos(x) + 5 x^4 \sin(x)$

Figure 3-1: DERIVE Results for the Derivatives of Some Products

If the result after expanding is the one you expected, try taking the derivative of several combinations of $f(x) \cdot g(x)$ using the table of derivatives you had DERIVE generate. In some cases you may have to do some algebraic simplifications. Are the results consistent with your conjecture?

The Laboratory Report

Report on the results of your investigation of the derivative of $f(x) \cdot g(x)$. Include evidence to support your conjecture for a formula for the derivative $(f \cdot g)'(x)$. Now extend your investigation to the derivative of the quotient $\left(\dfrac{f}{g}\right)'(x)$. To do this, essentially duplicate the above process, with the exception that you are looking at the quotients $\sin(x)/x$, . . . , $\sin(x)/x^5$.

Extended Credit

Consider the composition of functions $(f \circ g)(x)$. Look at $(\sin(x))^2, \ldots, (\sin(x))^5$ as well as $\sin(x^2), \ldots, \sin(x^5)$. What pattern do you see? Now try $\ln(\sin(x))$, $\sin(\ln(x))$, $\sqrt{\sin(x)}$, and $\sin(\sqrt{x})$. Does the same pattern emerge? What about a general rule for $(f \circ g)'(x)$?

Work Sheet

Laboratory #3

Using DERIVE fill in the following table:

$f(x)$	$\frac{d}{dx}\left(f(x)\right)$
x^n	$n\,x^{n-1}$
$\sqrt{x-1}$	
$\sin(x)$	
$\mathrm{atan}(x)$	
$\mathrm{asin}(x)$	
$\ln(x)$	

Choose some functions other than x^n from the above table as a choice for $f(x)$ and use DERIVE to determine the derivatives and fill in the following table

$f(x)$	$\frac{d}{dx}\left(f(x)\right)$	$\frac{d}{dx}\left(x\cdot f(x)\right)$	$\frac{d}{dx}\left(x^2\cdot f(x)\right)$	$\frac{d}{dx}\left(x^3\cdot f(x)\right)$	$\frac{d}{dx}\left(x^4\cdot f(x)\right)$	$\frac{d}{dx}\left(x^5\cdot f(x)\right)$

Based on this evidence, make a conjecture about the derivative of $x^n\cdot f(x)$ where n is a positive integer.

Extend this conjecture to the derivative of $f(x)\cdot g(x)$. Illustrate for two functions f and g.

Choose some functions other than x^n from the above table as a choice for $f(x)$. Use DERIVE to determine the derivatives and fill in the following table:

$f(x)$	$\frac{d}{dx}\left(f(x)\right)$	$\frac{d}{dx}\left(\frac{f(x)}{x}\right)$	$\frac{d}{dx}\left(\frac{f(x)}{x^2}\right)$	$\frac{d}{dx}\left(\frac{f(x)}{x^3}\right)$	$\frac{d}{dx}\left(\frac{f(x)}{x^4}\right)$	$\frac{d}{dx}\left(\frac{f(x)}{x^5}\right)$

Based on this evidence, make a conjecture about the derivative of $\frac{f(x)}{x^n}$ where n is a positive integer.

The above formula hides the general rule for the derivative of $\frac{f(x)}{g(x)}$. Choose two functions for f and g, where the choice for g is something other than x^n, and examine the derivative of this quotient. What is your conjecture for the derivative of a quotient.

Show that the derivative of $\frac{f(x)}{x^n}$ follows this general rule.

Laboratory #4

Determining the Graphs of Functions from Data

Introduction

In this lab we will consider the following problem: Given a small amount of data about a function, is it possible to identify the function and draw its graph? We will assume that we know the shape of the graph. As it turns out, this is a lot of information. We will be able to determine the function if we can identify a small number of points on the graph. This ability will prove to be very useful in situations where it is physically impossible to collect large amounts of data. For example, suppose we wish to construct a symmetric parabolic arch that must be a certain height and have its footers located at certain fixed positions. In this case we know only three points, but we need to know all about the arch in order to draw plans, order materials, and begin construction. You will see that you have enough information to completely describe the arch.

Throughout the lab you will be asked to display data on the graphics screen. This is done by authoring the data points using square brackets. For example, if you want to display the data given by the following table,

x	y
-1	2.1
0	1.5
1	$-.3$
2	-1.2
3	$-.5$
4	.2

author the following: $[[-1 , 2.1] , [0 , 1.5] , [1 , -.3] , [2 , -1.2] , [3 , -.5] , [4 , .2]]$. The square brackets are very important. Without them, you will receive an error message, and you will need to edit the data. The data is displayed in tabular form. With this data highlighted, press ⌷F1⌷ to change windows and "p" for Plot. To see all of the data points, zoom out on the x-axis. The data points appear as bright dots on the graph. This data is displayed in Figure 4 − 1. In this figure the data points have been manually enhanced to help distinguish them from the tick marks on the graph.

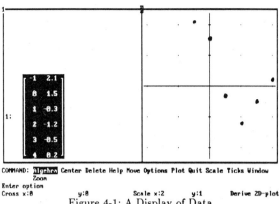

Figure 4-1: A Display of Data

Constructing the Graphs of Polynomials

Let's return to the problem of constructing a symmetric parabolic arch. Suppose we know that the arch is to be 100 feet high and that the two base supports are to be 500 feet apart. We begin by imposing a coordinate system on the space occupied by the arch. We will assign the coordinates [0,0] to the left-hand base of the arch and [500,0] to the right-hand base. This is an arbitrary choice, and you may decide to assign different coordinates to the base. For example, a strong case could be made for a choice of [-250,0] and [250,0]. This would have the peak of the arch on the y-axis. Using the original choice, the peak of the arch is at [250,100]. Enter the three points [[0,0],[500,0],[250,100]], using the DERIVE Author command, and plot the result.

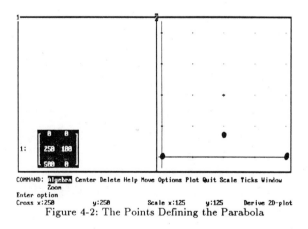

Figure 4-2: The Points Defining the Parabola

If these three points lie on a parabola, then there are constants a, b, and c such that the graph of $f(x) = ax^2 + bx + c$ passes through these points, i.e., $f(0) = 0$, $f(250) = 100$, and $f(500) = 0$.

These three conditions generate a system of three equations in three unknowns.

$$a \cdot 0 + b \cdot 0 + c = 0$$
$$a \cdot 250^2 + b \cdot 250 + c = 100$$
$$a \cdot 500^2 + b \cdot 500 + c = 0$$

We will use the Manage/Substitute option together with the soLve operator to find the values for a,b, and c that satisfy these equations. From the first equation, we see that c = 0. This means that we have a system of two equations in two unknowns.

$$a \cdot 250^2 + b \cdot 250 = 100$$
$$a \cdot 500^2 + b \cdot 500 = 0$$

This system can be easily solved by hand to yield $a = -\dfrac{1}{625}$ and $b = \dfrac{4}{5}$. The graph of f(x) for these values of a, b, and c is shown in Figure 4 − 3.

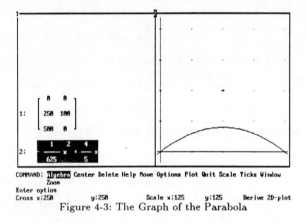

Figure 4-3: The Graph of the Parabola

Let's go back to the beginning and see how we can use DERIVE to find a, b, and c. Begin by authoring the expression $a \cdot x^2 + b \cdot x + c$. Assume that this is expression #1 on the DERIVE screen. Use the Manage/Substitute command to substitute x = 0 in this expression (simply press the Enter key when asked for values of a, b, c). This will be expression #2. Likewise, substitute the values x = 250 and x = 500 to obtain expressions #3 and #4 as shown in Figure 4 − 4.

The system of three equations in three unknowns can be entered by authoring the following expression

[#2 = 0, #3 = 100, #4 = 0].

Once again, the square brackets are essential. The screen display is the system enclosed in square brackets. This is also shown in Figure 4 − 4 as expression #6.

Pressing "L" for soLve gives the same solution for the system that we obtained by hand calculation (see expression #7 in Figure 4 − 4). We can now substitute these values for a, b, and c in expression #1 and plot the graph of f(x).

The method we used to solve for a, b, and c is independent of the data points given. If we are given three points $[x_1, y_1]$; $[x_2, y_2]$; and $[x_3, y_3]$, we simply repeat the above procedure substituting x_1, x_2, and x_3 for x. We then use the appropriate y values to enter a system of equations and solve them. Try this method to find the graph of a quadratic function that passes through the points [2, 20]; [- 3, 55]; and [1, 7]. Where is the vertex of this parabola located?

A word of caution: *It is not always possible to arbitrarily pick three points and find a quadratic that passes through the points.* The only time that we can succeed is when the three points produce a system of three independent, consistent equations in three unknowns. For example, replace the point [− 3,55] by [1,55] in the previous example. If we add the condition that none of the x-coordinates of the points are the same, can we always succeed? Will the resulting equation always be a true quadratic?

The method described above can be extended to polynomials of any degree. To illustrate this point, find a cubic function whose graph passes through the points [− 2, 4]; [− 1, 1]; [0,3]; and [1,3]. Use the methods of the calculus to locate the local maximum and minimum points on the graph of this function.

It is time to try a conjecture. In general, is it true that given n+1 points: $[x_1, y_1]$; $[x_2, y_2]$; · · ·; $[x_{n+1}, y_{n+1}]$ with $x_1 \neq x_2 \neq \cdots \neq x_{n+1}$, it is possible to find a polynomial of degree at most n passing through these points?

Including Information About the Derivative

In some situations it is impossible to find data at three distinct points on the graph. Our method of identifying a quadratic equation relied on having three equations in three unknowns to solve for the constants a, b, and c. Suppose we know the location of only two points on the graph, but happen to know the value of the derivative at one point. Is this enough information to uniquely identify the quadratic function passing through these points and having the prescribed behavior at one of the points? Suppose that we know $f(x_1) = y_1$, $f(x_2) = y_2$, and $f'(x) = p$. Since $f'(x) = 2ax + b$, we have the following set of three equations:

$$a \cdot x_1^2 + b \cdot x_1 + c = y_1$$

$$a \cdot x_2^2 + b \cdot x_2 + c = y_2$$

$$2 \cdot a \cdot x_1 + b = p$$

This system can be solved for a, b, and c as long as $x_1 \neq x_2$. For example, let's find the height of a symmetric parabolic arch that has a 500 foot base and, at the base, rises 3 feet per horizontal foot of change.

We begin by finding the expression for $f(x) = a \cdot x^2 + b \cdot x + c$ whose graph represents the arch. Let $(0,0)$ and $(0,500)$ be the coordinates of the base of the arch. We need to solve the following system:

$$a \cdot 0^2 + b \cdot 0 + c = 0$$
$$a \cdot 500^2 + b \cdot 500 + c = 0$$
$$2 \cdot a \cdot 0 + b = 3$$

The last equation uses the fact that $f'(0) = 3$. This system of equations is easily solved (with or without DERIVE) to yield $a = -\frac{3}{500}$, $b = 3$, and $c = 0$. Using the Manage/Substitute command, we create an expression for $f(x)$ that has these values. Highlight this expression and press "c" for the Calculus menu and "d" for Differentiate. The proper response for the next three requests is to press the Enter key. Press the "l" for soLve, and you see that the maximum is at $x = 250$ as you expected. Substituting this value into the expression for $f(x)$ gives us a height of 375 feet for the arch.

Use the technique illustrated in the above paragraph to find a parabola that passes through the point $(1,6)$ with a slope of 12 and that also passes through the point $(0, -6)$. What happened in this case? Note the slope of the line joining the two points.

The Laboratory Report

Write a report to assist an engineer who is planning to build a bridge that will span a 300- foot gap in a mountain road across a ravine. On one side of the gap the upward grade in the road is 7% (slope .07). On the other side of the ravine the road is 20 feet above the level of that on the first side and the grade is a 5% downward grade (slope $-.05$). The engineer wants to build a bridge that follows a parabolic arc and smoothly joins with the existing road. Show that this is not possible. Find a function that will meet the engineer's specifications. Write your report in clear, concise English. You may assume that the engineer has had calculus. Carefully explain why the parabolic shape will not work and why you chose the solution you are presenting.

Work Sheet

Laboratory #4

Your lab report is to deal with finding the shape of the bridge described on the previous page. The first exercises on this sheet are warm-ups to get **you** in shape for solving the problem of your lab report. These exercises parallel the material in the text of Laboratory 4.

1. Write the four equations in four unknowns needed to find the cubic curve

$$ax^3 + bx^2 + cx + d = f(x) = y$$

   that passes through the four points $(-2, 4)$; $(-1, 1)$; $(0, 3)$; and $(1, 3)$.

   Use DERIVE to solve this system for a, b, c, and d. Graph the resulting curve.

2. Using the information given about the function and its derivative, find a, b, and c such that

$$f(x) = ax^2 + bx + c$$

   and $f(-2) = 23$, $f(1) = 2$, and $f'(-2) = 22$.

3. Show that if $f(x) = ax^2 + bx + c,$ it is not possible to find a, b, and c such that $f(0) = 0,$ $f(300) = 20,$ $f'(0) = .07,$ and $f'(300) = -.05.$

4. Use the techniques developed in this lab to find a cubic function that will meet the above specifications.

Laboratory #5

Using the Derivative to Approximate the Value of a Function

Introduction

Suppose you were asked to find the area (in square feet) of a square measuring 2 feet 2 inches on a side. Can you give the answer quickly? Well, perhaps you can, but even if you can't, you should be able to give a quick estimate for the area. How? By using the derivative and linear approximation. You know that a square of length x feet on a side, has area, $A(x) = x^2$. You also know that the area of a square 2 feet on a side is 4 square feet and that the square whose area you are trying to estimate has sides of length $2\frac{1}{6}$ feet. How can we put all this knowledge together to come up with an answer?

Sometimes half of the battle is to carefully rephrase the question. We will rephrase it as a question about the area function, $A(x)$. Given that $A(x) = x^2$ and $A(2) = 4$, what is the value of $A(1+\frac{1}{6})$? This restatement of the question should place you in mind of finding the value of $A(x+h)$ for some small increment, h, given the value of $A(x)$.

Recall the definition of the derivative:

$$A'(x) = \lim_{h \to 0} \frac{A(x+h) - A(x)}{h}$$

From this definition, we see that we can approximate the difference $A(x+h) - A(x)$ by $h \cdot A'(x)$, i.e.,

$$A(x+h) - A(x) \approx h \cdot A(h)$$

or,

$$A(x+h) \approx A(x) + h \cdot A'(x) . \tag{1}$$

Returning to the original problem,

$$A(2+\tfrac{1}{6}) \approx A(2) + \tfrac{1}{6} \cdot A'(2)$$

$$A(2+\tfrac{1}{6}) \approx 4 + \tfrac{1}{6} \cdot (2 \cdot 2) = 4\tfrac{2}{3}$$

Thus, we say that the area of the square is approximately $4\frac{2}{3}$ square feet. Use your pocket calculator to compare this approximation with the actual value for the area of the square.

Equation (1) above is the basis of the linear approximation technique. Given a function, $f(x)$, whose value is known at some $x = x_0$, it is possible to approximate the value of f at

$x = x_0 + h$ close to x, if we also know the value of $f'(x)$ at $x = x_0$, using the formula,

$$f(x_0+h) \approx f(x_0) + h \cdot f'(x_0). \qquad (2)$$

The accuracy of the approximation depends upon the size of h and the shape of the function, $f(x)$ at the point $(x_0, f(x_0))$. Figure $5-1$ shows the graphs of two different functions, $f(x)$, and the linear approximations, $4 + f'(2) \cdot (x-2)$, to the functions at the point $(2,4)$. Note that the line in the graph on the left appears to coincide with the graph of $f(x)$ more than the one on the right.

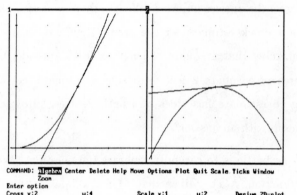

COMMAND: Algebra Center Delete Help Move Options Plot Quit Scale Ticks Window
Zoom
Enter option
Cross x:2   y:4   Scale x:1   y:2   Derive 2D-plot

Figure 5-1: Graphs of $f(x)$ and Linear Approximations to $f(x)$ at $(2,4)$

Extending the Linear Approximation Technique

As a population of bacteria grows, competition for resources causes the growth rate of the population to decline. The following data was gathered concerning the rate of growth of a bacterial population. The time, t, is given in days.

t	P'(t)
0.00	1.000
0.20	0.962
0.50	0.800
0.75	0.640
1.00	0.500
1.25	0.390
1.50	0.308
2.00	0.200
3.00	0.100
4.00	0.059

Given this data, taken over nonuniform time intervals, we would like to determine the ultimate size of the bacterial population. We begin by graphing the data as shown in Figure $5-2$.

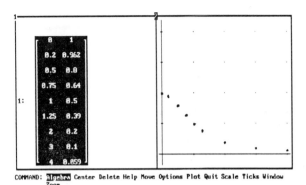

Figure 5-2: The Growth Rate of the Bacterial Population

From the above graph, it is clear that the growth rate of the population is approaching zero and that if we can approximate the function value at $x = 4$, we will have a good idea of the ultimate population size of the bacterial colony. Suppose that we also know that the initial size of the colony was $1 \, cm^2$. We will successively approximate the value of the function at $t = .2,.5,.75,1,1.25,1.5,2,3,4$ using linear approximation techniques, i.e.,

$$P(t_{i+1}) \ = \ P(t_i) \ + \ (\text{rate of change})_i \cdot (t_{i+1} - t_i) \tag{3}$$

Note that the above equation contains $(\text{rate of change})_i$ instead of $P'(t_i)$. The reason for this is that we have more information about the way in which the function is changing. For example, in the time interval from $t = .2$ to $t = .5$, the rate changes from $.962$ to $.8$. Thus, we may assume that the average rate of change in the interval is the average of these rates, or $\frac{1}{2} \cdot (.962 + .8)$, or $.881$.

We will now use DERIVE to assist us in approximating the size of the population at $t = 4$. Begin by entering the data from the table on page 38. We will need to refer to individual entries or elements of that table. Thus, we need to learn some terminology. To refer to the element in row 3, column 1, we author "element(#1,3,1)." Simplifying this expression yields $\frac{1}{2}$ or $.5$. If we author the expression "element(#1,6,2)," we see after Simplifying, $\frac{39}{100}$ or $.39$. This is the element in row 6, column 2. Our goal is to create an expression that can be easily be evaluated several times using the Manage/Substitute combination. Thus, we author the rather awkward expression

$$P \ + \ \frac{(\text{element}(\#1,i+1,2) + \text{element}(\#1,i,2))}{2} * (\text{element}(\#1,i+1,1) - \text{element}(\#1,i,1))$$

This expression is nothing more than equation (3) written in DERIVE syntax that refers to the table.

Begin the approximation process. Substitute P=1 and i=1 into the above expression and obtain a new value, P = 1.962, for P at t = .2 (use the approXimate operator). Substitute this value for P and 2 for i to obtain 1.4605 as the new value for P. Continue in this fashion to obtain an approximation for P at t = 4. Create a new table of values of t and P(t) and Plot the data. The graph is displayed as Figure 5 − 3.

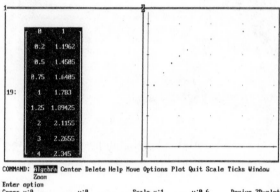

Figure 5-3: The Actual Size of the Bacterial Population

The Laboratory Report

The following data is from an acceleration test of a 1989 base Porsche 911 Carrera 4.[1,2]

Time(sec)	Velocity(mph)
0	0
1.8	30
2.8	40
3.9	50
5.1	60
6.8	70
8.6	80
10.5	90
13.0	100
16.2	110
19.7	120
25.6	130

---

[1] *Car and Driver 35(1989), p.43*

[2] This example was suggested by Professors Elgin Johnston and Jerry Matthews of Iowa State University.

Write a report that compares the acceleration of this automobile to that of a top secret test car that is being developed in an atmosphere of strict industrial secrecy. The only information that you can obtain is that the prototype of the new car accelerated from 0 to 130 mph in $\frac{3}{4}$ mile.

Since you cannot develop a table similar to the one given in *Car and Driver* for the Porsche, you will need to determine how far the Porsche traveled during the *Car and Driver* test. In writing your report be careful that you use meaningful units to determine the distance.

A graph of the Porsche test data is given Figure 5 − 4.

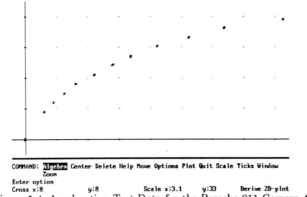

COMMAND: Algebra Center Delete Help Move Options Plot Quit Scale Ticks Window
Zoom
Enter option
Cross x:8   y:8   Scale x:3.1   y:33   Derive 2D-plot

Figure 5-4: Acceleration Test Data for the Porsche 911 Carrera 4

DERIVE 2.0

Using the Looping Structure in DERIVE

There is a certain value to doing the manual procedure of the previous sections. It makes the steps of the linear approximation very clear. The newly obtained value of the function is used as the starting point for the next approximation and the data on the derivative is updated to the next value of the independent variable. All of these steps are handled by you in the Manage/Substitute process. However, if you had to do this for any sizable number of steps, you would soon find that the process becomes rather tedious. It would, in this case, be preferable to write a program to do the process.

DERIVE version 2.0 and above has this facility using a procedure called Iterates. The format for this procedure is

Iterates(<expression>,<substitution variable>, <initial value>,<# of iterations>).

The operation of the procedure is that the <expression> will be evaluated at the <initial value> and the <substitution variable> will be replaced by the result of the evaluation. This process will be repeated <# of iterations> times, each time using the most recent value of the variable For example, approximating the expression

Iterates(cos(x), x, 0, 40)

will generate a vector of 40 values the last 5 of which are .739085 . In this vector the first element is the cosine of 0. Each succeeding element is the cosine of its predecessor. Since the sequence eventually settles on the value .739085, this value is one such that, up to a precision of 6 decimal places, the value is the same as its cosine.

The Iterates operator seems, at first glance, to have limited applicability. Its function is to perform successive compositions of the <expression> at the <initial value>. This is a useful operation, but it is not immediately obvious how the operator can be used in our situation. The answer lies in the fact that we can combine this operator with the Vector structure to do some programming. A simple illustration is writing a DERIVE statement that will generate a list of the integers from 1 to 10 and their squares. To do this we create a vector that will act much like the statements in a looping process, i.e., have one element that will update the counting variable and another element that is the square of the first. Name the <substitution variable> v, and initialize it to [1, 1] (the integer 1 and its square). During each successive step we will need to update the first coordinate by one and replace the second by the square of the new first coordinate. Thus, the new value of v will be [element(v,1)+1, (element(v,1)+1)^2]. We will generate 9 values beyond the initial value of v. Authoring,

Iterates([element(v,1)+1, (element(v,1)+1)^2], v, [1, 1], 9)

Yields the table

$$
\begin{bmatrix}
1 & 1 \\
2 & 4 \\
3 & 9 \\
4 & 16 \\
5 & 25 \\
6 & 36 \\
7 & 49 \\
8 & 64 \\
9 & 81 \\
10 & 100
\end{bmatrix}
$$

Thus, DERIVE can perform an action that is much like executing a programmed FOR loop.

The process we are doing is a more advanced than the one used to illustrate the workings of the Iterates procedure given above. We are approximating the values of a function from an initial value of the function and several values of its derivative. This means that we need to update two items, the value of the function and the values of the derivative used to make the next approximation. Thus, we need to use a vector as our <substitution variable>. This vector will contain the number of the approximation, so that the appropriate values of the derivative can be taken from the appropriate array. The second element of the vector will be the value of the function.

To construct the Iterates expression to perform our task, we author the vector for the values of the derivative and construct the DERIVE equivalent of the linear approximation formula (3) found on page 39. This is where we began the Manage/Substitute process. Assume that expression #M contains that linear approximation formula. We will also assume that the vector we will be creating as our <substitution variable> is called, v. We begin by substituting in expression M the following values to create expression #M+1

for i (the element # from the derivative vector): element$(v, 1)$
for P (the value of the function): element$(v, 2)$

This makes an extremely long expression, but we are not concerned with that at the moment. Now we are ready to construct the Iterates expression for the biological data example.

$$\text{iterates([element}(v,1) + 1, \#M+1], v, [1,1], 9)$$

This expression generates a table of data that contains the value for i and the updated population size that is similar to the one highlighted in Figure $5-3$. Note that we started with initial values of i $= 1$ and P $= 1$. Each successive entry is the update of these two variables. For example, the second entry is [2, 1.962] and the third is [3, 1.4605], and so on until we have i=10 and the second entry is the value for P at the 10th entry in the table (the size of the population when t=4).

For your laboratory report, you will be using the twelve data values for the Porsche. Thus, you will start with the vector [1, 0] and have 11 iterations.

Work Sheet

Laboratory #5

Recall that the instantaneous velocity of the Porsche is the derivative of its position at that time. Thus, if

$$s(t) = \text{position of the Porsche at time t}$$

then

$$s'(t) = v(t) = \text{velocity of the Porsche at time t}$$

1. Since the data for the Porsche is given in seconds and miles per hour, what constant will convert miles per hour to feet per second?

2. Write the linear approximation expression to approximate the position of the Porsche in feet after t seconds.

3. Write the DERIVE equivalent of this linear approximation expression.

4. Using the above, fill in the following table

Time	i	Position
0	1	0
1.8	2	
2.8	3	
3.9	4	
5.1	5	
6.8	6	
8.6	7	
10.5	8	
13.0	9	
16.2	10	
19.7	11	
25.6	12	

Laboratory #6
Graphing Revisited

Introduction

We begin with an allegory

Three blind people were asked to examine an elephant and tell what they had learned about the animal. They were told that an elephant is a large animal. The first person examined the head of the elephant and told the others that an elephant is a large bird with big and mighty wings. "Oh, no!" said the second, who had examined the trunk of the beast. "An elephant is like a giant cobra. It is long and thick, possessing great power." "You are both wrong," said the third, who was stationed near the rear legs. "An elephant is extremely tall and has wrinkly, baggy skin like an aged person." Together they concluded "This is a strange beast, indeed."

An examination of the individual conclusions of each person shows that, given what each had chosen to describe and the limited means that each had to explore the elephant, the three individuals all did a reasonably good job. The problem was one of piecing together the information they had gathered.

What do the two graphs in Figure 6 − 1 have in common?

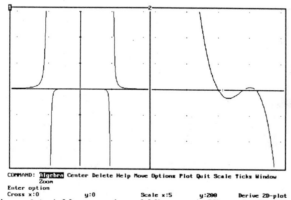

COMMAND: **Algebra** Center Delete Help Move Options Plot Quit Scale Ticks Window
Zoom
Enter option
Cross x:0   y:0   Scale x:5   y:200   Derive 2D-plot

Figure 6-1: A Macroscopic and Microscopic View of a Function

The answer is that they are both graphs or, more accurately, parts of graphs of the same function. Each has information the other lacks, and neither have all of the information necessary to describe the function. This is the problem of a limited window and limited display capabilities.

Although computer graphics provide us with an excellent tool for examining the local behavior of functions, they do not necessarily provide a complete description of the function under consideration. Like all tools, it is best when in the hands of a skilled crafter. The function shown in Figure 6 − 1 is

$$f(x) \; = \; \frac{2x^6 - 4x^5 - 9x^4 + 36x^2 + 81x - 162}{x^6 - 75x^4 + 1875x^2 - 15625} \; .$$

We will use DERIVE to take several snapshots of the function and then show how the calculus provides us with a consistent strategy to investigate the global as well as local behavior of any function. We will also demonstrate how DERIVE can be a useful computational tool in the latter investigation.

Before we start, adjust the precision and mode of operation for DERIVE. After entering DERIVE and setting up an algebra and a graphics window, press "o" for Options while in the graphics window. You will see a new menu that allows us to set the Mode and Precision. The default setting for the Mode is Exact. This means that DERIVE does arithmetic by working with rational numbers and radicals. Change the mode to Approximate by pressing "a". This switches the way in which DERIVE does arithmetic. It now acts like a calculator and uses numerical algorithms. Now the cursor is in the Precision column. This is the number of significant digits for the calculator mode. Because DERIVE's default accuracy of 6 digits is not enought to reveal all of the detail we will explore, change the Precision to "10", and press the Enter key. We are now ready to begin our exploration. Begin by drawing graph shown on the left side of Figure 6 − 1. Set the x-scale to 5 and the y-scale to 200.

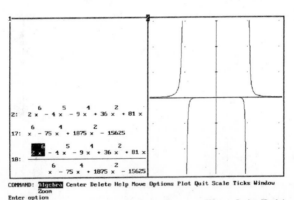

Figure 6-2: The Graph of the Function Drawn with an Expanded y-Scale

In this view of the function, it appears that the function has vertical asymptotes at x = ± 5 and a horizontal asymptote along the x-axis (y = 0). The function looks absolutely flat from x = − 3 to x = 3. Of course, with a y-scale of 200, it is impossible to distinguish anything but the most extreme change in the y-direction. The only change that we see is at the vertical asymptotes.

Readjusting the y-scale to 20, we suspect that the horizontal asymptote for the function is something other than 0. It also appears that the asymptote along the negative x-axis might be different from that along the positive axis.

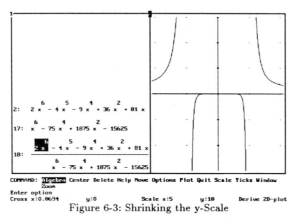

Figure 6-3: Shrinking the y-Scale

Changing the x-scale to 40 quickly dispels any illusion that there may be different horizontal asymptotes for negative and positive x-values. Digitizing the point on the graph at x = ±80 illustrates that the line y = 2 is a horizontal asymptote. This point can be illustrated more vividly by changing the y-scale to 5, or even 2.

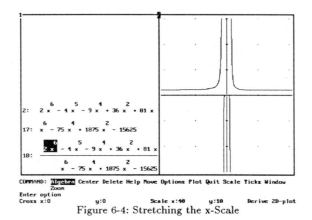

Figure 6-4: Stretching the x-Scale

But what is happening between x = −5 and x = 5? In each of Figures 6−2, 6−3, and 6−4, the function appears to be absolutely flat on the interval [-4,4]. Given the nature of the function definition, we know that this cannot be. (Why?) We begin by changing the x-scale to 2. In this graph we see only the center part of the function, but the graph still appears to lie flat on the x-axis. Zooming in on the y-axis will eventually produce a picture similar to that in Figure 6−5. Although the y-scale is rather small, this is behavior more typical of what we should expect. There is however, another flat spot near x = 2.

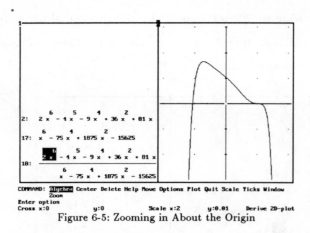

Figure 6-5: Zooming in About the Origin

Move the cross hair to x = 2, and change the x-scale to .1 and the y-scale to .000005, and note the result.

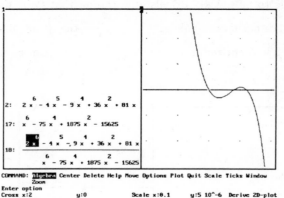

Figure 6-6: Greatly Increasing the Magnification About (2,0)

Our investigation of this function had us travel through eight orders of magnitude. If we piece together our observations, we can give a reasonable description of the function's behavior. In the next section we will present a unified procedure for examining the behavior of the function.

A Systematic Procedure Using Calculus

In this section we outline a procedure for investigating the general behavior of a function. Our example function will be the f(x) of the previous section.

There are four main steps to the procedure we will follow.

1. Locate the vertical asymptotes.

2. Locate the horizontal asymptotes.

3. Determine the critical points for the function.

4. Examine the behavior of the function between the critical points.

In each of the steps we will use DERIVE to assist in our analysis.

Return to the graphics window and close it, leaving only the algebra window on the screen. This is done by pressing the "w" and "c" for close. This gives us a complete screen for displaying our results.

Since we are dealing with a reduced rational function, we begin searching for vertical asymptotes at those values of x for which the denominator is zero and the numerator is nonzero. To do this highlight the denominator of the expression for f(x) and press "f" for Factor. In response to the request for the amount of factoring, press "d" for raDical. and then press the Enter key. The resulting screen display is

$$\frac{2x^6 - 4x^5 - 9x^4 + 36x^2 + 81x - 162}{(x-5)^3(x+5)^3}$$

Thus, we see that the denominator is zero when $x = \pm 5$. Substituting these values for x into the numerator of f(x) by using the Manage/Substitute option yields 14268 and 38458, respectively. Thus, the roots of the denominator are not roots of the numerator, and we suspect that f(x) may have vertical asymptotes at $x = \pm 5$.

We examine the behavior of the function at these suspected asymptotes. In particular, we will examine the following limits

$$\lim_{x \to -5^-} f(x), \qquad \lim_{x \to -5^+} f(x), \qquad \lim_{x \to 5^-} f(x), \qquad \text{and} \qquad \lim_{x \to 5^+} f(x).$$

To evaluate the first limit, highlight the expression for f(x). Now press "c" for Calculus and "l" for Limit. In response to the request for the limit variable type "x" and press the Enter key. The next request will be for a point. Type -5 and then press the Tab key. This moves you over to the direction field. Since we are looking for the limit as x approaches -5 from the left, type "l" for Left and then press Ctrl and Enter keys together. This last action causes DERIVE to immediately simplify the result without first displaying the symbolic representation and forcing you to request the simplification. In essence, it saves a step. After a brief wait, the symbol ∞ is displayed on the screen.

Repeat the above steps, making one change. After Tabbing to the direction field, press "r" for Right. In this case, the result is $-\infty$. This is the symbolic description of the behavior of the function as it approaches the line $x = -5$ from the left and right corresponding to the display in Figure $6-2$. Complete this symbolic description by evaluating the two limits at $x = 5$. This completes the investigation of the function behavior at the vertical asymptotes.

To determine the horizontal asymptotes for the graph of f, we evaluate the two limits,

$$\lim_{x \to -\infty} f(x) \qquad \text{and} \qquad \lim_{x \to \infty} f(x) .$$

These limits describe the behavior of the function as x becomes "large." The process is exactly the same as the above process, except that when entering the point, type -inf or inf, respectively. The limit as $x \to -\infty$ will, of course be from the right. As $x \to \infty$, the limit will be from the left. In both cases the answer is 2. Thus, the line $y = 2$ is a horizontal asymptote.

We now search for critical points of the function. These are points where $f'(x) = 0$ or fails to exist. Highlighting the expression for f(x) and using the Differentiate operator on the Calculus menu, we see

$$f'(x) \; = \; \frac{4x^6 - 282x^5 + 500x^4 + 756x^3 - 405x^2 - 828x - 2025}{(x^4 - 50x^2 + 625)(x^2 - 25)^2}$$

The denominator of this expression is simply $(x - 5)^4 (x + 5)^4$. A quick check shows that 5 is not a root of the numerator. Thus, $f'(x)$ fails to exist at $x = \pm 5$, the values for x that are not in the domain of f. Our first attempt to find the zeros of the numerator is to factor it. After waiting a few minutes, we see that DERIVE is making no progress towards factoring this expression. To stop this effort and try something more profitable, press the Esc key.

Recall that we set DERIVE in Approximate mode with a Precision of 10. After highlighting the numerator of the expression for $f'(x)$, press "l" for soLve. In Exact mode DERIVE would try to find the roots of the specified expression with no further information required. In its present mode, it needs to have a search interval specified. Our knowledge of the algebra of polynomials tells us that no root of this polynomial can exceed ± 2025, begin by specifying the lower bound on the root as -2025 and the upper bound as -5. This will search the axis to the left of the first vertical asymptote. After about a half minute, the message "No Solutions Found" appears in the lower left-hand portion of the screen. We conclude that there are no critical points to the left of the vertical asymptote at $x = -5$.

Now repeat the process with -5 as the lower bound and 5 as the upper bound. This time the screen displays $x = -1.439754785$ in the algebra window. This value corresponds to the x-coordinate of the local maximum shown in Figure $6-5$. We will give analytic evidence that the graph of f does indeed have a maximum for this value of x after we finish looking for critical points. In order to determine that there are no additional critial points in this interval, also search from -5 to -1.43.

Continue the search with a lower bound of -1.43 and an upper bound of 5. The next value given is $x = 2.032371164$, which corresponds to the minimum point in Figure $6-6$. Changing bounds to 2.04 and 5 yields a value of x = 2.103018239. No additional roots are found in the interval 2.11 to 5. So far we have located three values of x for which f'(x) ≈ 0. However, the numerator of the expression for f'(x) is a sixth degree polynomial. This means that there must be an even number of real roots. Either one of the roots has multiplicity 2, or there is at least one more value of x for which f'(x) = 0.

Setting the lower bound at 5 and the upper bound for a solution at 2025 yields a result of x = 68.6390868. If we center the graph on x = 68 and y = 2 with an x-scale of 20 and a y-scale of .03, we can see some evidence of a local minimum point in the vicinity of x = 68. Continuing the search in the interval 68.65 to 2025 yields no more critical points.

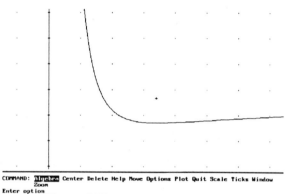

COMMAND: Algebra Center Delete Help Move Options Plot Quit Scale Ticks Window
Zoom
Enter option
Cross x:68.0555   y:2.0001   Scale x:20   y:0.03   Derive 2D-plot

Figure 6-7: A Minimum Point That Was Overlooked When Only Viewing Graphs

We conclude our discussion by finding f''(x) and substituting the values of x found above. The results are

$$f''(-1.439754785) < 0, \quad f''(2.032371164) > 0, \quad f''(2.103018239) < 0, \quad f''(68.6390868) > 0.$$

We can make the following analysis of f at this point:

> The graph of f has a horizontal asymptote at y = 2 and vertical asymptotes at x = ±5. The function has no critical points in the interval $(-\infty, -5)$. Checking the values of the first and second derivative at an arbitrary point in this interval allows us to conclude that the graph is increasing, concave upward in the interval $(-\infty, -5)$. The graph increases from x = -5 to -1.43 . . . , where it has a local maximum. It then decreases to local minimum at x = 2.03 . . . , and then increases to a local maximum at x = 2.10. . . . The graph decreases from that point along the vertical asymptote at x = 5. Decreasing from the vertical asymptote at x = 5, the graph attains another local minimum at x = 68.639 . . . , and then increases, approaching the horizontal asymptote from below. Further

examination of the function values at the local maximum and minimum points will show that the changes in function values at these points are very subtle.

Missing from this analysis is any discussion of the inflection points on the graph. This topic is left for you to investigate.

The Laboratory Report

Conduct an analytic examination of the behavior of the function

$$f(x) = \frac{x^4 - 2x^3 - 7x^2 + 20x - 12}{x^4 - 2x^3 - 27x^2 - 52x - 28}$$

and describe its behavior for all x. Include the appropriate graphs in your report to support and explain your analysis. Use DERIVE to assist you in your analysis. Carefully explain each of your conclusions and how you arrived at them.

Work Sheet

Laboratory #6

Given the function

$$f(x) = \frac{x^4 - 2x^3 - 7x^2 + 20x - 12}{x^4 - 2x^3 - 27x^2 - 52x - 28}$$

use DERIVE to supply the following information

Horizontal asymptote(s):

Vertical asymptote(s):

$f'(x) =$

Critical points:

$f''(x) =$

Local maxima, minima:

Graph of f(x):

Laboratory #7
You and Derive — A Problem-Solving Duo
(Applied Optimization Problems)

Introduction

One of the temptations when given a powerful system such as DERIVE is to believe that there is no need to learn calculus , or much other mathematics, because it is all "on the disk." No idea could be sillier. Such a thought is akin to thinking that there is no need to plan a trip because the automobile does all of the work of getting us there. DERIVE is billed as a "Mathematical Assistant." This means that just as the automobile can extend our physical horizons, DERIVE and other CAS packages can expand our problem-solving horizons. We are not as limited by the need to do difficult technical manipulations as we were in the past. We are free to consider the truly interesting and challenging aspects of problem-solving . In short, you and DERIVE are a "dynamic problem-solving duo." But remember, you are the problem-solving Batman and DERIVE is Robin.

Let's look at the process of solving an applied optimization problem (max-min, or calculus word problem). After listing the steps in the process, we will sort out those tasks that are either best done or must be done by you, and those that are best done by DERIVE.

1. Carefully read the problem and determine what information you have been given. What is being optimized, i.e., what is your objective or purpose for solving the problem? What are the variables and constants that influence how you attain the objective?

2. Construct a mathematical model of the situation described in the problem statement, using the information learned in step 1. Determine if calculus is the appropriate tool to use for solving the problem.

3. After constructing the model and determining that calculus is the appropriate tool, locate the critical points of the function and determine the concavity of the function at each of these points.

4. Evaluate the function at those points that are candidates for the optimum value (each of the maxima or minima, depending on the type of problem you are solving, and the end points of the interval over which the objective function is defined. Choose the point that gives the "best" value for the function.

5. Analyze your result and interpret its meaning.

Steps 3 and 4 of the above process are rather mechanical and are the areas where DERIVE can increase your productivity. The other steps are what you do best. They require you to analyze and synthesize the material you are given. You have to be able to identify the crucial pieces of information given in the problem statement and find the relationships that exist between these pieces. Much research is being conducted on programming computers to do this kind of analysis, but so far they have only been able to operate in a very limited domain. Many researchers are convinced that a computer program will never be able to outperform a good problem solver.

An Example Problem and Solution

A ship is patrolling enemy waters and needs to be very discrete. It will not penetrate the waters to a point off the coast closer than 6 miles from shore. While on a patrol, the ship experiences some difficulty and must drop anchor at the point where the difficulty occurred. This is at the 6-mile limit.

Fortunately, the seas are calm and there is a contact on the shore 10 miles from the point on shore opposite where the ship is anchored. The ship's difficulty has also caused its radio to be ineffective. Therefore it is decided to send a courier to deliver a message for help to the contact. The only boat they have is a small lifeboat. They choose a crew member who is capable of rowing at a rate of 2 mph in the calm sea and running along the coast road at a rate of 6 mph for the distance that needs to be covered.

Since time is of the essence, find the optimal landing point on the coast for the crew member to end the sea-journey part of the mission and begin running. Although it is not exactly the case, assume that the coast road is a straight line.

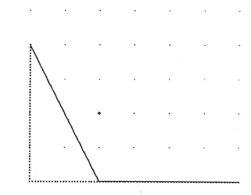

Figure 7-1: A Possible Path to the Contact

The first, and sometimes most useful, thing that we can do is to draw a picture of the situation described in the statement of the problem. This is exactly what we do in Figure 7-1. The vertical dotted line represents the distance from the ship to the shore (6 miles). The horizontal dotted line represents the shore road. The solid lines represent a possible path for the crew member to follow. This picture will aid in the analysis of the problem.

Restating the problem in terms of the picture, "At what point along the shore road should the crew member land in order to minimize the total travel time required to reach the contact?" From this restatement, we can determine that the objective is to minimize time traveled. The trip includes sea travel (rowing the boat) and land travel (running along the road). Thus,

$$\text{Travel Time} = \text{Time Rowing} + \text{Time Running} \qquad (1)$$

From an introductory physics course, we know

$$\text{Time Rowing} = \frac{\text{Distance to Row}}{\text{Rowing Speed}}$$

$$\text{Time Running} = \frac{\text{Distance to Run}}{\text{Running Speed}}$$

We know that the crew member has a rowing speed of 2 mph and running speed of 6 mph. Thus, we need to find a way to describe the "Distance to Row" and "Distance to Run." Fortunately, Pythagoras comes to our rescue.

What we need to know is the point on the coast road to land the boat. As is the usual custom, call this point "x". This x is the distance of the horizontal dotted line in Figure $7-1$. Since the entire line is 10 miles long, the remainder of the line is: $10-x$ and that x lies between 0 and 10. This is the distance the crew member must run. What about rowing? This is where we need Pythagoras. The vertical dotted line is 6 miles long, so the hypotenuse of the triangle (the path for rowing) is $\sqrt{x^2 + 36}$ miles long. Thus,

$$\text{Time Rowing} = \frac{\sqrt{x^2 + 36}}{2}$$

$$\text{Time Running} = \frac{(10 - x)}{6}$$

The total time of the journey is thus a function of the variable x.

$$T(x) = \text{Travel Time} = \frac{\sqrt{x^2 + 36}}{2} + \frac{(10 - x)}{6}$$

We are now ready to enlist the aid of DERIVE. The really hard work is finished, and we settle down to what amounts to collecting the data. We begin by authoring the expression

$$T(x) := \sqrt{(36 + x\char94 2)/2} + (10 - x)/6$$

The screen display is shown as expression #1 of Figure 7 − 2. The complete DERIVE session is shown in Figure 7 − 2. We simply had DERIVE compute the derivative of T(x) and solve for the values of x for which $T'(x) = 0$. In this case there is only one such value and as the graph of T(x) shows, it is a minimum point. Alternatively, we could have tested this point using the second derivative test.

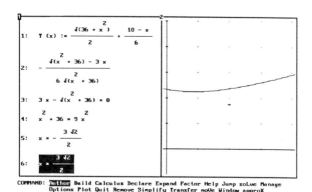

Figure 7-2: The DERIVE Session to Find the Landing Point

Expressions #3 and #4 are interesting to examine. When we set $T'(x) = 0$ and solved, the version of DERIVE we are using did not give an answer. It merely simplified the expression to

$$3x - \sqrt{x^2 + 36} = 0$$

At this point we must intervene and do a very little algebra. Rewrite the expression as

$$x^2 + 36 = 9x^2$$

DERIVE can very easily solve this expression and give the result: $x = \dfrac{3\sqrt{2}}{2} \approx 2.12132$. After we determine that this is not a spurious solution introduced by squaring both sides of the original equation, highlight the expression for $T'(x)$ and find its derivative. Substitute the value 2.12132 for x in the expression for $T''(x)$ to determine that the graph of T(x) has a local minimum for this value of x. Substitute $x = 0$, $x = \dfrac{3\sqrt{2}}{2}$, and $x = 10$ for x in T(x) and choose the value that gives the minimum. In this case the absolute minimum occurs at the local minimum when $x = \dfrac{3\sqrt{2}}{2}$.

Extending the Process − Adjusting Parameters

There are times when the parameters in the discussion may be variable, and policy decisions may be in force that tell us for what value of the independent variable the optimum for the objective function must occur. This presents us with a slightly different type of problem. For what values of the parameters will the objective function obtain its optimum value for a given value of the independent variable? DERIVE's symbolic capabilities are a great help in the solution of this problem. The problem-solving procedure is much the same as the above, except that we must extend the analysis step.

In order to continue with our present model, we make the following addition to the problem

The crew member reaches the contact and after delivering the distress message, recalculates the running speed. The contact informs the crew member that there is a supply of motors of various horsepowers that can be attached to the life-boat eliminating the need to run. Since a faster motor will make more noise and increase the chances of being caught, the crew member decides to use the smallest motor that will propel the boat at a speed to make an all water-return trip optimal. At this stage the running speed and boat speed are unknown. What relationship between these two speeds makes an all-water return optimal?

In this revised problem we know that the optimum value for $T(x)$ is to occur when $x = 10$. What is unknown at that moment are the values for the speed of the boat, say b, and the running speed, say r. The objective function, in this case, is

$$T(x) \; = \; \frac{\sqrt{x^2 + 36}}{b} \; + \; \frac{(10 - x)}{r}$$

Differentiate this expression with respect to x and solve for x. Once again, you will have to assist DERIVE. The optimum value for x in terms of b and r is (see expression #7 of Figure $7 - 3$)

$$x \; = \; 6 \sqrt{- \frac{1}{b^2 - r^2}} \; |b|$$

Note, this expression only makes sense if $0 < b < r$. Obviously, if $b > r$, no analysis is needed; an all-water trip is optimal.

Now impose the requirement that $x = 10$ by substituting this value for x in the above expression. This results in expression #8 of Figure $7 - 6$. We modify this to the quadratic

$$100(r^2 - b^2) \; = \; 36 \, b^2$$

DERIVE finds two solutions for this equation, and we choose the positive one for obvious reasons. Thus the relationship between the boat speed and the rowing speed that will make the all-water return trip optimal is $b = \frac{5\sqrt{34}}{34} \, r \; \approx 0.857432 \, r$. In other words, if the boat speed is approximately 86% of the crew member's running speed, an all-water trip is optimal in this situation.

Bureaucracies being what they are, it is decided that every captain who visits this spot should be prepared for a similar situation. Thus, a chart is prepared to help the captain decide, based on the running speeds of the crew members, how fast the life boat must be propelled. A "policy space chart" is designed. This chart is displayed in Figure $7-3$. The horizontal axis represents the variable b and the vertical axis, the variable r. The Captain merely determines r and moves across to the line. Any

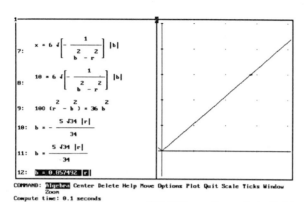

Figure 7-3: The DERIVE Session for the Extended Problem and the 'Policy Space'

value of b that lies to the right of the line will guarantee that an all-water trip is optimal. The value of the b coordinate on the line is the minimum value for b that will suffice. Of course, if the ship is further down the coast, or the seas are rough, a different chart will have to be constructed.

The Laboratory Report

You and your partners have been commissioned by the Interstate Commerce Commission to prepare a report on the following situation.

After conducting an extensive study of Interstate Trucking, the Interstate Commerce Commission has concluded that the primary variable expenses for any over-the-road freight hauler are the wages for the driver and the cost of fuel. The study concluded that the maintenance and replacement costs for vehicles, although considerable, did not vary significantly from carrier to carrier. On the other hand, fuel costs and wages did vary. The Commission is interested in knowing if there is an optimal wage-to-fuel-cost ratio that will make most haulers encourage their drivers to abide by a policy of traveling 55 mph or less as dictated by national legislation.

Assume that your group has gathered the following data. Under ideal conditions, an interstate freight hauler gets 6 miles per gallon of fuel. This mileage is greatly affected by the average speed of the vehicle and its weight. In general, the miles per gallon of fuel decrease by .2 for each increase of 10,000 pounds of weight over the optimal 25,000 pounds that the truck and freight weigh. Also, the miles per gallon of fuel decrease by .1 mile per gallon for each mile per hour that the truck averages over a speed of 45 mph. This information can be used to create an expression for the effect of weight and average speed on the cost per run of fuel.

At the present time the national average for diesel fuel is $1.25 per gallon, and the average over-the-road wage for a driver is $15.00 per hour. If the average weight of a loaded truck is 75,000 lbs, what is the optimal average speed for a truck under these conditions?

Assuming that the average weight will not vary, determine a fuel-to-wages ratio that will keep a speed of 55 mph or less optimal for the freight haulers. Include in your report a policy space for this problem and instructions on its use for the ICC. In particular, explain how the policy space can be used to determine the minimum cost for fuel that will keep the optimal speed limit at 55 mph or below if the average over-the-road wage of the drivers is known.

Work Sheet

Laboratory #7

1. Starting from the equation

Cost of a trip = driver's wages + fuel cost for the trip

construct a mathematical model for use in your policy decision.

driver's wages/mile =

$$\text{fuel cost/mile} = \frac{\text{cost/gallon}}{\text{miles/gallon}}$$

The function you will be optimizing:

2. Use the above model to find the optimum average speed given the present conditions mentioned in the discussion of the problem on the preceeding pages.

3. Fuel prices have risen drastically to $2.75 a gallon. What is the new optimum average speed under this condition if all other conditions stay the same, i.e., driver's wages and truck weight? What can be done to bring the optimum average speed closer to 55 mph without changing the truck weight?

Laboratory #8

Summation Notation and Some Sums

Introduction

Suppose you were instructed to evaluate the sum

$$1 + \frac{1}{2} + \frac{1}{4} + \frac{1}{8} + \frac{1}{16} + \frac{1}{32} + \frac{1}{64} + \frac{1}{128} + \frac{1}{256} + \frac{1}{512} + \frac{1}{1024}$$

It is possible to use DERIVE to author the sum exactly as it appears above and simplify the result to

obtain an answer of $\frac{2047}{1024}$. The problem with this procedure is that it is tedious, prone to error, and

requires much more effort than is necessary to do the task.

Even the most casual examination of the above sum reveals that each term is half of the preceding term, i.e., if we label the terms with subscripts $0, 1, \ldots, 10$, we see

$$\text{term}_i = \frac{1}{2} \cdot \text{term}_{i-1} \qquad i = 1, 2, \ldots, 10$$

Furthermore, the first term in the sum (the one with subscript 0) is 1. Thus,

$$\text{term}_1 = \frac{1}{2} \cdot 1 = \frac{1}{2^1}; \qquad \text{term}_2 = \frac{1}{2} \cdot \frac{1}{2^1} = \frac{1}{2^2}; \qquad \cdots \qquad ; \qquad \text{term}_{10} = \frac{1}{2} \cdot \text{term}_9 = \frac{1}{2^{10}} .$$

Also, noting that $1 = \frac{1}{2^0}$ reveals that for the above sum,

$$\text{term}_i = \frac{1}{2^i} \qquad \text{for } i = 0, 1, \ldots, 10 .$$

We have found a formula for the ith term of the sum in terms of its location within the sum (its index).

Mathematical notation provides us with a convenient way to take advantage of a sum that involves such terms. If a_i is the ith term of a sum for $i = 0, \ldots, n$, then the notation,

$$\sum_{i=0}^{n} a_i$$

denotes the sum of the terms, a_i, for $i = 0, \ldots, n$. In other words, it is shorthand for

$$a_0 + a_1 + \cdots + a_n$$

For our example, the appropriate notation is

$$\sum_{i=0}^{10} \frac{1}{2^i}$$

Since this notation is very common in the development of the calculus, DERIVE provides an easy method to enter and evaluate such sums.

Begin by authoring the general, i^{th} term of the sum: $1/2^i$. The two-dimensional display of the term appears on the screen and is highlighted. Press "c" for calculus , and in the Calculus menu press "s" for Sum. Since you are summing the highlighted expression, press the Enter key. The sum variable is i. In response to the Limit question, enter 0 for the lower limit and 10 for the upper limit. When you press the Enter key, the display will be similar to the expression at the top of the page. Simplifying this expression gives the result of $\frac{2047}{1024}$ that was mentioned earlier.

Not only is this procedure easier, it is more general. For example, we can easily change the limits of summation. Since DERIVE works symbolically, we can enter n as the upper limit for the sum and obtain the general formula

$$\sum_{i=0}^{n} \frac{1}{2^i} = 2 - 2^{-n} .$$

This is the familiar sum of a geometric series with ratio $\frac{1}{2}$.

Practice with this process on some other geometric series with ratios $\frac{1}{3}$, $\frac{1}{5}$, 2, 3, 5 and 1. This last result may surprise you and seem to be out of character with the rest of the results.

Sums Involving Powers of Consecutive Integers

There is a story told of the mathematician Carl Friedrich Gauss that when he was in the equivalent of kindergarten he was bothering his teacher with embarrassing questions. The teacher, in an effort to silence Gauss, told him to add all of the numbers from 1 to 100. Gauss immediately replied that the sum was 5,050. When asked to explain how he obtained the result so quickly, Gauss replied that it was quite simple, "100 plus 1 is 101; 99 plus 2 is 101; 98 plus 3 is 101; and so it continues until 50 plus 51. Thus, the result is 50 · 101, or 5,050."

We extend Gauss' result to the sum of the first n positive integers, $\sum_{i=1}^{n} i$. Note that Gauss' technique involved summing pairs at opposite ends of the summation. Thus, we write the sum forwards and backwards. We then add these two equivalent expressions for the sum, term by term, and solve for the value.

$$\text{Sum} = 1 + 2 + \cdots + (n-1) + n$$
$$\text{Sum} = n + (n-1) + \cdots + 2 + 1$$
$$2\cdot\text{Sum} = (n+1)+(n+1)+\cdots+(n+1)+(n+1) = n \cdot (n+1)$$

or,

$$\text{Sum} = \frac{n \cdot (n+1)}{2} \; .$$

You can quickly check the result with Gauss' observation, by substituting 100 for n.

Check DERIVE's ability by applying the summation procedure with $\text{term}_i = i$, lower limit: 1, and upper limit: n. Now, extend the above demonstration to a formula for evaluating the sum,

$$\sum_{i=m}^{n} i \; .$$

Check your result by author ing the sum in DERIVE and simplifying the resulting expression. It may require some algebraic manipulation to show that you and DERIVE have the same answer.

What about the summation of powers of consecutive integers. The simple, direct reasoning used by Gauss does not seem to work here. For example, what is the formula for

$$\sum_{i=1}^{n} i^2 \; ?$$

If we construct the expression in DERIVE and simplify, we see a value of

$$\frac{n \cdot (n+1) \cdot (2n+1)}{6}$$

This result is not at all obvious. How did such an expression appear?

We will demonstrate the correctness of the formula by comparing the sums of squares of consecutive integers with the sums of consecutive integers. Construct and fill in the following table.

n	$c_n = \sum_{i=1}^{n} i$	$s_n = \sum_{i=1}^{n} i^2$	$\frac{s_n}{c_n}$
1	1	1	1
2	3	5	$\frac{5}{3}$
3	6	14	
4			
5			
$\vdots$			
n	$\frac{n(n+1)}{2}$	?	

After completing the table you should easily spot a pattern for $\frac{s_n}{c_n}$. Multiplying this ratio by the value of c_n will give you the result that was displayed by DERIVE.

The Laboratory Report

Present and discuss your derivation for the value of the sum, $\displaystyle\sum_{i=1}^{n} i^2$. Also, apply a similar method to the evaluation of the following sums.

$$\sum_{i=1}^{n} i^3$$

In each of these cases, compare the result of your derivation with the result obtained using DERIVE.

Work Sheet

Laboratory #8

Using the DERIVE summation operator, fill in the following table

n	$c_n = \sum\limits_{i=1}^{n} i$	$s_n = \sum\limits_{i=1}^{n} i^2$	$\dfrac{s_n}{c_n}$
1			
2			
3			
4			
5			
6			
7			

Conjecture: $\dfrac{s_n}{c_n} =$

Evaluation of $\sum\limits_{i=1}^{n} i^2 = \left(\dfrac{s_n}{c_n}\right) \cdot c_n = \left(\dfrac{s_n}{c_n}\right)\left(\dfrac{n(n+1)}{2}\right) =$

Repeat the above for $\sum\limits_{i=1}^{n} i^3$.

n	$c_n = \sum\limits_{i=1}^{n} i$	$s_n = \sum\limits_{i=1}^{n} i^3$	$\dfrac{s_n}{c_n}$
1			
2			
3			
4			
5			
6			
7			

Laboratory #9
The Definition of the Definite Integral

Introduction

How do you find the the area of an irregularly shaped surface? The ancient Greeks considered this problem and came up with a clever solution that is the foundation for the theory of the definite integral. What the ancients did was to slice the surface into thin strips. Each of these strips had a shape that was very close to that of a rectangle (see Figure $9-1$). They then calculated the areas of each of the rectangles , added these areas together, and had a good approximation of the area of the surface. It is obvious that thinner strips give a better approximation of the overall area. This observation ignores the effects of errors in addition caused by the increased number of terms to be added.

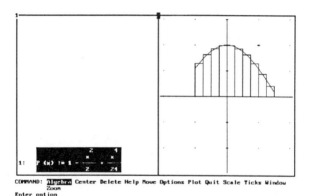

Figure 9-1: A Surface and a Rectangular Polygon Approximating the Surface

We will consider surfaces that can be described as the area lying on the plane bounded on one side by the curve $y = f(x)$, on the other by the interval [a,b], and on the sides by the lines $x = a$ and $x = b$, respectively. We use the convention that the signed area lying below the axis is negative. Figure $9-1$ depicts a representative surface. If we assume that the value of the function at the midpoint of each strip represents the average height of the strip, then the expression

$$f(m_i){\cdot}(x_i - x_{i-1})$$

approximates the area of the strip whose base is the interval $[x_{i-1}, x_i]$. The area of the surface can then be approximated by summing these expressions for each $i = 1, \ldots, n$.

$$\text{Area} = \sum_{i=1}^{n} f(m_i){\cdot}(x_i - x_{i-1})$$

Continuing with our assumptions, we assume that the interval [a,b] is sliced into rectangles whose bases all have the same width. This width is the length, $b - a$, of the interval divided by n, the number of slices. Thus,

$$\text{Area} = \sum_{i=1}^{n} f(m_i) \cdot \frac{(b-a)}{n} = \frac{(b-a)}{n} \cdot \sum_{i=1}^{n} f(m_i)$$

It remains to determine the midpoint, m_i, of the interval $[x_{i-1}, x_i]$. For the first interval, the midpoint is halfway between x_0 and x_1, i. e.,

$$m_1 = x_0 + \tfrac{1}{2}(x_1 - x_0) = a + \tfrac{1}{2} \cdot \frac{(b-a)}{n}$$

Each successive midpoint is an interval's width away from the midpoint of the previous interval. Thus,

$$m_i = m_{i-1} + \frac{(b-a)}{n} = m_{i-2} + \frac{2(b-a)}{n} = \ldots$$

$$\ldots = m_1 + \frac{(i-1)\cdot(b-a)}{n} = a + \tfrac{1}{2} \cdot \frac{(b-a)}{n} + \frac{(i-1)\cdot(b-a)}{n}$$

The final expression for the estimate of the area is

$$\text{Area} \approx \frac{(b-a)}{n} \sum_{i=1}^{n} f\left(a + \frac{(b-a)}{2n} + \frac{(i-1)\cdot(b-a)}{n}\right) \qquad (1)$$

If we use the above to estimate the area of the curve in Figure $9-1$ with $a = -1$, $b = 1.5$, $n = 10$, and $f(x) = 1 - \frac{x^2}{2} + \frac{x^4}{24}$, the area of the surface is approximated as 1.84707. This is a rather good approximation to the actual area of 1.84245.

A closer examination of equation (1) reveals that the process of approximating the area of the surface is one of accumulating the values of the function, $f(x)$, that makes up the upper boundary of the surface; i.e., we choose one value of the function in the interval $[x_{i-1}, x_i]$ for each $i = 1, \ldots$, $n-1$, sum these values, average them (divide by the number of intervals), and multiply by the width of the interval. It is this process of accumulation that is of interest to us. We will examine the relationship of the accumulated values of the function to the function itself. Our observations may prove to be surprising. Since the calculations would be tedious if done by hand, we begin our investigation by defining an expression for the accumulated values of a function in DERIVE.

Using DERIVE to Evaluate the Sum

We begin by alerting the DERIVE system that we plan to use the letter f to stand for a function. To do this we press "d" for Declare and respond to the prompts in the following way. The responses to the requests are given in quotes.

> DECLARE: "f" (for Function)
>
> DECLARE FUNCTION name: "f" (for the name of our function)
>
> DECLARE FUNCTION value: "<Enter>" (we will supply expressions later)
>
> DECLARE FUNCTION variable: "x" (the standard independent variable)
>
> DECLARE FUNCTION variable: "<Enter>" (there is only one independent variable)

The expression $F(x):=$ appears on the screen.

Having reserved the name, f, for a function name, we author the expression for the value of the function at the ith midpoint, m_i .

$$f(a + (b-a)/(2n) + (i-1)(b-a)/n)$$

The next step is to sum these values over $i = 1, \ldots, n$. Press "c" for calculus and "s" for Sum. We must, once again, give a series of responses. Press the Enter key for all responses except the second.

> SUM variable: "i" (DERIVE had chosen "a" because of its position in the alphabet)

The screen display for this series of responses is

$$\sum_{i=1}^{n} F\left(a + \frac{(b-a)}{2n} + \frac{(i-1)(b-a)}{n}\right)$$

Check your screen to ensure that the display matches the above expression. If it does not match, remove your expression and repeat the above sequence of steps.

We are now ready to enter the equivalent of expression (1). We declare a function S(a,b,n)

$$S(a,b,n) := \frac{(b-a)}{n} \sum_{i=1}^{n} F\left(a + \frac{(b-a)}{2n} + \frac{(i-1)(b-a)}{n}\right) \tag{2}$$

This is done in a manner similar to the way in which we defined f as a function. There are two differences. The first is our response to "DECLARE FUNCTION value:" . The correct response is "(b-a)/n*#3" (assuming that #3 is the number of the sum expression). The second difference is that we declare three function variables, a, b, n. Each of these is entered in succession, one at a time. The resulting screen display is the same as expression (2).

Check this expression by Declaring F(x) to be $1 - \frac{x^2}{2} + \frac{x^4}{24}$ by entering this expression in response to the "DECLARE FUNCTION value:" request. Next, author the expression $S(-1,1.5,10)$ and press "x" for approXimate. The expression is evaluated as 1.84707.

The Relationship of the Accumulated Values of f(x) to the Function, f(x)

The function S(a,b,n) is a function of three variables. Begin by fixing a = 0, and letting b = x. Then S(0,x,n) is the sum of the accumulated values of f over the interval [0,x]. If we arbitrarily choose n = 10, then S(0,x,10) is a function that approximates the area of the surface bounded by the function f(x) and the x-axis lying above (or below) the interval [0,x].

Recall that our current function definition is $f(x) = 1 - \frac{x^2}{2} + \frac{x^4}{24}$. Author the expression S(0,x,10). Plot this expression to obtain a plot similar to Figure 9 − 2. If you are using a split screen display, let the x-scale be 2. The plotting process will be rather slow, but the point is made that S(0,x,10) is certainly a function of x. Return to the Algebra window and simplify the expression S(0,x,10). Note that the simplified result has a fifth-degree, a third-degree, and a first-degree term. If it is necessary to Plot S(0,x,10) again use the simplified form. It plots much more quickly.

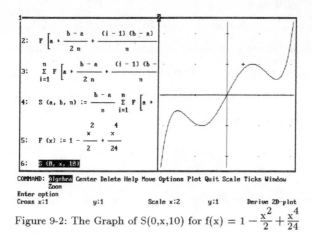

Figure 9-2: The Graph of S(0,x,10) for $f(x) = 1 - \frac{x^2}{2} + \frac{x^4}{24}$

But the primary question remains, "What is the relationship of S(0,x,10) to the function, f(x), that is used to define S?" In an attempt to answer this question, a display of the graphs of the two functions in the same plot is given in Figure 9 − 3.

This plot is crucial to our understanding of the relationship of f and S. If you need help in distinguishing between the two functions, label the graphs in Figure 9 − 3. Complete the following table:

Sign of f(x)	Behavior of S(0,x,10)
Negative	
Zero	
Positive	

The table and the graphs in Figure $9-3$ should convince you of the relationship between the two functions. This relationship is not exact as can be seen by differentiating the simplified version of S(0,x,10).

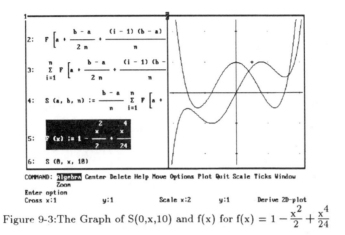

Figure 9-3:The Graph of S(0,x,10) and f(x) for $f(x) = 1 - \dfrac{x^2}{2} + \dfrac{x^4}{24}$

Check your conjecture of the relationship between f and S by drawing the graphs of f(x) and the simplified version of S(0,x,10) for the following choices of f(x). This can be done easily in three steps by declaring the new function, f(x), authoring S(0,x,10), and simplifying the result.

1. $f(x) = x^3 - x + 1$

2. $f(x) = \sin(x)$

3. $f(x) = |x|$

4. $f(x) = \dfrac{1}{x^2 + 1}$ (It will take some time for DERIVE to simplify S(0, x, 10))

5. $f(x) = \sin(x^2)$

The Laboratory Report

Your report is to be a statement and defense of your conjecture about the relationship between the function f(x) and the function S, obtained by accumulating the values of f over the interval [0,x]. Include in your report the graphs that you generated as part of your investigation. Based on your observations, what relationship do you expect between the following function and f(x)?

$$S(x) = \lim_{n \to \infty} \frac{x}{n} \cdot \sum_{i=1}^{n} f\left(\frac{x}{2n} + \frac{(k-1)x}{n} \right)$$

Test your answer by declaring some simple polynomial functions and taking the limit of S(0,x,n).

Work Sheet

Laboratory #9

Enter the expression for S(a, b, n) into DERIVE. Don't forget to declare a function, f(x), prior to entering this expression.

For each of the following functions, draw the graph of f(x) and $S(0, x, 10)$. After graphing each pair of f(x) and $S(0, x, 10)$, fill in a table similar to the one given below.

1. $f(x) = \sin(x)$

2. $f(x) = \dfrac{1}{x^2 + 1}$

3. $f(x) = x^3 - x + 1$

4. $f(x) = \text{abs}(x)$

5. $f(x) = \sin(x^2)$

Sign of f(x)	Behavior of $S(0, x, 10)$
Negative	
Zero	
Positive	

Laboratory #10

An Application of the Definite Integral:
Estimating Heating Costs

Introduction

Let's think about averaging a group of numbers, for example, the test scores for your calculus class on the last exam. To make matters easy, assume there are only 10 students in the class. The scores are given in the array

$$A := [52, 77, 68, 97, 88, 92, 85, 78, 83, 94] \ .$$

To calculate the average, merely add all of the scores and divide the sum by 10. The mathematical notation for this process is

$$\frac{\sum_{i=1}^{10} A_i}{10} = \text{Average of A} \ .$$

If there are more than 10 students in the class, you simply replace the 10 in the above by n, the number of students. Of course, the array, A, will have n entries.

This idea works fine for a finite set of numbers, but suppose we want to find an average value of a function f(x), defined on an interval [a, b]? One way to approach this problem is to divide the interval [a, b] into n subintervals of equal width and randomly sample the function in each subinterval. The average of these n function values can then be used as an estimate of the average value of the function, i.e.,

$$\text{Average}_{[a, b]} \text{ of } f \approx \frac{\sum_{i=1}^{n} f(\xi_i)}{n} \ , \text{ where } \xi_i \in [x_{i-1}, x_i] \ , \text{ the } i^{th} \text{ interval.}$$

The more function values we sample and average, the more meaningful the estimate of the average. Thus, we define

$$\text{Average}_{[a, b]} \text{ of } f = \lim_{n \to \infty} \frac{\sum_{i=1}^{n} f(\xi_i)}{n}$$

The right-hand side of the above expression looks very familiar. If we multiply both sides of the above by the length of the interval, we see

$$(b\text{-}a) \cdot \text{Average}_{[a, b]} \text{ of } f = \lim_{n \to \infty} \frac{(b\text{-}a) \cdot \sum_{i=1}^{n} f(\xi_i)}{n} = \lim_{n \to \infty} \frac{(b-a)}{n} \sum_{i=1}^{n} f(\xi_i) = \int_{a}^{b} f(x)\,dx$$

Dividing both the left- and right-hand sides of the equation by $(b - a)$ we arrive at the definition,

$$\text{Average}_{[a, b]} \text{ of } f = \frac{1}{(b - a)} \int_{a}^{b} f(x)\,dx$$

On the other hand, if we look at the left- and right-hand sides of the equation at the bottom of page 73, we see that the integral of f over the interval [a, b] is the limit of an accumulation of the values of f on this interval. It is this idea that we will use in the following sections.

Determining A Delivery Schedule for Heating Oil

A heating oil firm gets a new customer on January 11. The firm's policy is to deliver a full tank of oil (250 gal) and then refill the tank 10 days later. By noting the daily temperature fluctuations and the amount of oil required by the customer on the first fill-up, the firm can work out a delivery schedule for the driver who will be servicing the customer. The firm wants to schedule the deliveries so that the customer's tank never drops below 50 gallons and also so the driver is not making any unnecessary trips to merely top off the customer's tank. The delivery schedule is to be made up through the end of March.

The heating-industry standard for determining the heating requirements of a site is the number of gallons of fuel used per Heating Degree Day. This method of estimation has proven to be extremely accurate. Heating Degree Days are determined by the following formula.

$$65 - \frac{(\text{Daily High Temp} + \text{Daily Low Temp})}{2}$$

Only days for which Heating Degree Days is positive will contribute to the heating oil consumption. Table $10 - 1$ is a record of the temperature data for the 10-day period between deliveries.

When the oil delivery is made on January 21, the customer requires 167 gallons of fuel. This, coupled with the temperature data, gives the firm a way of figuring the heating requirements of the customer in terms of gallons of fuel per Heating Degree Day (HDD).

$$\text{gal/HDD} = \frac{\text{Fuel Consumed During Time Period}}{\text{Total Heating Degree Days in Period}} = \frac{167}{376} \approx 0.444$$

From this information, the firm then determines the next delivery date, so that

$$0.444 \cdot \text{Accumulated HDD} = 200 \,.$$

We must now find a way to determine the accumulated HDD during the time period in question. To do this, we will need to find a function for the HDD and use the integral to determine the accumulation.

Date	Day of Year	Daily Temperature High	Low	Degree Days
January 11	11	41	20	34.5
January 12	12	38	22	35
January 13	13	35	17	39
January 14	14	36	20	37
January 15	15	35	10	42.5
January 16	16	36	18	38
January 17	17	34	20	38
January 18	18	38	18	38
January 19	19	36	22	36
January 20	20	35	19	38

Table 10-1: Temperature Data for $1/11 - 1/20$

To find a function that estimates the HDD, we consider the average yearly temperature data given in Table $10 - 2$ and look at a graphical display of this data given in Figure $10 - 1$.

Date	Day of Year	Daily Temperature High	Low	Degree Days
January 15	15	36	22	36
February 15	46	39	23	34
March 15	74	49	31	25
April 15	105	63	41	13
May 15	135	73	51	3
June 15	166	82	61	-6.5
July 15	196	86	65	-10.5
August 15	227	85	65	-10
September 15	258	78	57	2.5
October 15	288	66	45	9.5
November 15	319	53	36	20.5
December 15	350	40	26	32

Table 10-2: Average Temperature Data for the Year

The graph of the data in the table has the appearance of a sine curve. Certainly, it should come as no surprise that temperature data is periodic. We expect the temperatures during certain times of the year to be the same as at the same time during preceding years. If the temperatures vary greatly from the expected norm, it is a newsworthy event.

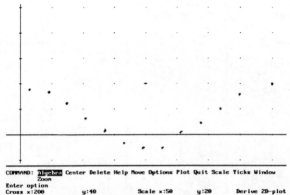

Figure 10-1: A Graphic Display of the Data From Table 10-2

Thus, we are going to guess that the function that describes the average HDD function is a sine function with period 365, i.e.,

$$HDD(t) = A + B \cdot \sin\left(\frac{2\pi}{365}(t - C)\right).$$

We need to determine the values for A, B, and C to complete our analysis. Examining the data and the graph of the data leads us to the following analysis

The maximum value for HDD is 36, and its minimum is −10.5. Thus, the amplitude of the sine function is approximately 23. If this is the case, then the graph of the function will be centered about the line y = 13. The curve crosses this line in an upward sweep on approximately day 300. Therefore, our initial estimate for HDD is

$$HDD(t) = 13 + 23\sin\left(\frac{2\pi}{365}(t - 300)\right)$$

A graph of this function is displayed as the top graph in Figure 10 − 2.

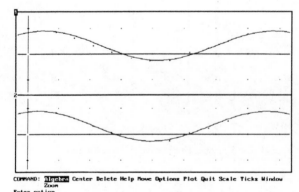

Figure 10-2: Two Attempts to Find Continuous Curves to Fit the Data

This graph follows the general shape of the data, but could be improved by shifting it to the left and slightly downward. Experimenting with some modifications of the function, the following function seems to give a fairly close fit to the data in Table $10 - 2$.

$$\text{HDD(t)} = 12.5 + 23\sin\left(\frac{2\pi}{365}(t-290)\right)$$

The graph of this function and the original data[3] is shown in Figure $10 - 2$.

In order to determine the next delivery date for the customer, we must find an X such that

$$\int_{21}^{X} [12.5 + 23\sin\left(\frac{2\pi}{365}(t-290)\right)]\ dt = \frac{200}{.444} = 450.45$$

This equation takes into account the facts that the last delivery was made on day 21, the estimated fuel consumption is .444 gallon per Heating Degree Day, and the firms desires to make a delivery of approximately 200 gallons of fuel.

Using DERIVE in Exact mode, the above equation can be simplified and then approximated to
$$-1336.1\cos(0.0172142\,X + 1.29106) + 12.5\,X - 371.628 = 450.45$$
Then shifting to Approximate mode with six significant digits, we can solve the equation in the range from 21 to 100 to yield the solution, $X = 33.8594$. Thus, the next delivery should be on day 34, or Feb. 3. To find the next delivery date, solve the equation

$$\int_{34}^{X} [12.5 + 23\sin\left(\frac{2\pi}{365}(t-290)\right)]\ dt = 450.45$$

and continue in this manner, substituting the new delivery date for the lower limit of the integral, until X exceeds 90, i.e., the next delivery date is past March 31. The delivery schedule for the customer is

February 3

February 16

March 3

March 20

April 13

Verify these dates. Remember that to evaluate the integral, DERIVE must be in Exact or Mixed mode. To solve the resulting equation, DERIVE must be in Approximate mode. Thus it will be

---

[3]The data for the tables and graphs in this report were obtained from the *1990 Susquehanna Valley Weather Almanac* published by WGAL-TV8 in Lancaster, PA, Joe Calhoun, Meteorologist.

necessary for you to change modes using the Options/Precision commands during the calculation to determine each date.

The Laboratory Report

The customer calls the heating oil firm and wants to know when oil will be delivered so that the neighbors will know when to expect the delivery truck. You share the above schedule with the customer, who does not understand why there is not a regular delivery date. Write a letter to the customer explaining how the delivery dates are determined.

After talking with the customer, you learn that the average daily temperatures from mid-February until the end of March will be 10 degrees warmer than usual. How will this affect the above delivery schedule? Include the updated schedule in your letter to the customer.

Finally, the customer has asked to be placed on the budget plan, i.e., to pay a fixed monthly cost based on estimated fuel consumption. If fuel oil averages $1.75 per gallon, determine the customer's monthly bill based on the estimate of yearly consumption. Note negative Heating Degree Days do not contribute to fuel consumption.

Work Sheet

Laboratory #10

Using DERIVE, draw the graph of the average temperature data for the year and the function

$$12.5 \ + \ 23 \sin\!\left(\frac{2\pi}{365}(t - 290)\right)$$

Given that oil is delivered on April 13 (day 103), find the date for the next oil delivery. Be careful! You may need to consider the summer months when the customer is not using any heating oil.

Assuming that the temperatures from Feb 16 to March 10 are 10 degrees warmer than usual, and that the company keeps to its policy of delivering as close to 200 gallons as possible, when will the next delivery after February 16 take place?

On the average, how many gallons of fuel oil can the customer expect to use each year?

Laboratory #11

An Application of the Definite Integral:
Underwater Navigation

Introduction

What do the following three objects have in common?

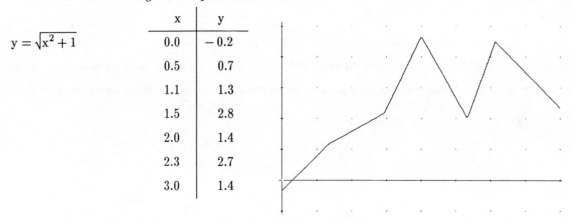

$y = \sqrt{x^2 + 1}$

x	y
0.0	− 0.2
0.5	0.7
1.1	1.3
1.5	2.8
2.0	1.4
2.3	2.7
3.0	1.4

Figure 11-1: Three Objects That Share a Common Property

The answer is that each of the above are valid representations of a functional relationship. The representation on the left is an equation. This is the type of function representation we have come to expect. It is the type that is most often encountered in mathematics books. The representation in the middle is a table. Note that for each given value of x, there is only one corresponding value of y. This is the definition of a function. The fact that we do not have a continuum of x-values does not violate the definition. The representation on the right is a graph. Note that the graph, although quite jagged, does not violate the definition of a function. It passes the vertical line test. For each value of x, there is only one value of y. If you look closely, you will note that the graph is constructed of straight lines joining the points given in the table. This is sometimes called the *line graph* for the data in the table.

Now, since all of the three objects above are representations of a function, say f(x), how can we evaluate the expression

$$\int_a^b f(x)\, dx \quad ?$$

In the case of $f(x) = \sqrt{x^2 + 1}$ the answer lies within the realm of a standard calculus course. Virtually every student who has passed Calculus 1 can do this integral. DERIVE can evaluate this integral in practically no time at all. The other two cases require more explanation.

The case of the table was essentially handled in Lab 4. In that lab we chose to assume that the integral has a value equal to the average value of the function over the interval times the width of the interval. We evaluated the integral as

$$\int_a^b f(x)\,dx \;=\; \sum_{i=1}^{n} (x_i - x_{i-1}) \frac{y_i + y_{i-1}}{2}$$

where (x_i, y_i) is the ith entry in a table that has $n+1$ entries numbered 0 to n.

To handle the case of a function represented as a graph the very definition of the definite integral tells us that

$$\int_a^b f(x)\,dx \;=\; A_+ \;-\; A_-$$

where A_+ is the area lying above the x-axis and below the curve, and A_- is the area lying above the curve and below the x-axis. There is a practical problem with this definition of the definite integral. It is not always possible to determine the exact values for the areas. In these cases we must estimate the area. This can be done by drawing the graph on graph paper and counting squares. Another way is to choose several points in the interval and make a table of the function values at those points. Now that you have a table, the integral is approximated using the summation method shown in the above paragraph. This is the method that we will use in the following example.

If you need to be refreshed on the method of determining function values from the values of the derivative, refer to Lab 4.

Review of Straight-Line Motion

For the following review, we will use the following notation:

$$s(t) \;=\; \text{Position of a body at time t}$$
$$v(t) \;=\; \text{Instantaneous velocity of the body at time t}$$
$$a(t) \;=\; \text{Acceleration of the body at time t.}$$

The relationship between these three functions is

$$v(t) \;=\; \frac{d}{dt} s(t)$$

and

$$a(t) \;=\; \frac{d}{dt} v(t) \;=\; \frac{d^2}{dt^2} s(t)$$

Thus, by the Fundamental Theorem of Calculus,

$$v(t) = \int_0^t a(x)\,dx + v(0) \qquad \text{and} \qquad s(t) = \int_0^t v(x)\,dx + s(0).$$

What this means in practice is that if it is possible to determine values for the acceleration of an object over some time interval ending at the present time, and you know the initial position and initial velocity of the object, you can determine the present position of the object.

Let's look at a small example. The following is hypothetical acceleration data for a rocket during its first 5 minutes of flight.

Time (minutes)	Acceleration (ft/sec^2)
0	44
1	73.3
2	88
3	80
4	70
5	65

We will also assume that the rocket had 0 initial velocity and was at position 0. Note that we are dealing with a uniform time interval. This means that we can adjust for the interval width following the calculation of the sums.

Calculating the average over each interval and accumulating the sum, we have the following table.

Time (minutes)	Velocity (strange units)
0	0
1	58.65
2	139.3
3	223.3
4	298.3
5	365.8

The top line of the table is the initial velocity data. To convert this table to miles per hour, multiply the results by $\frac{60}{5280}$. Why? Verify these results using DERIVE.

Repeating this procedure on the above table will give the distance of the rocket from the launch site. In this table the results were multiplied by the appropriate constant to convert them to miles from the launch site. What is this constant?

Time (minutes)	Distance (miles)
0	0
1	19.99
2	87.48
3	211.09
4	388.91
5	615.31

Thus, the acceleration data coupled with the initial velocity and initial position were sufficient to help us calculate the position of the rocket at prescribed times. Do your results agree with these?

Underwater Navigation

Recall that when you are riding in a car and the car accelerates too quickly, you are thrown back in your seat. That is why cars now have head supports to protect your head and neck in the event of a rear-end collision or an abrupt reversal of direction of the forward force. Also, if a driver applies the brakes too quickly, you are thrown forward. Hence, seat belts.

In each of these cases you are feeling a real force on your body. Newton's laws can explain this phenomenon. In particular,

$$\text{Force} = \text{mass}*\text{acceleration} .$$

Since the mass is remaining constant, the force you feel on your body is due to the acceleration of the vehicle transporting your body. This principle and the approximation method you used above are put to use in the navigation of submarines and other vehicles that have no visual way to compute their position.

It is possible to mount a stylus on a very sensitive spring and draw a piece of graph paper beneath it at a uniform rate. If the vehicle accelerates or decelerates (negative acceleration similar to putting on the brakes), the needle moves upward or downward on the graph paper as a result. Thus, the ship's computers have a record of the ship's acceleration. Assuming that the initial position and velocity of the ship are known, the computers can accurately figure the present position. Of course, as in the case of the rocket computation, the computer must do two numerical integrations.

In the situation shown in Figure $11-2$, a submarine starts at a known position on the ocean bottom and heads due east. The recording device is pointed along the axis of the submarine and records the strip shown in the Figure. Each horizontal tick mark on the x-axis is 7.5 units. Thus, the strip represents an hour voyage. Each tick on the vertical axis represents 2 ft/sec^2.

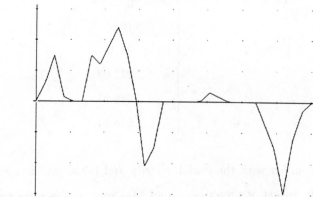

Figure 11-2: A Record of the Sub's Acceleration During a One-Hour Submersion

The Laboratory Report

Explain how the graph in Figure $11-2$ can be used to determine the position of the submarine at the end of the one-hour undersea voyage. Collect at least 30 data points from the above graph and use this data to determine the position of the submarine.

Work Sheet

Laboratory #11

Using the graph from Figure $11-2$, collect data on the acceleration of the submarine for every 2 minutes and use the techniques developed in the manual to complete the following table.

Time(min)	Acceleration	Velocity	Position
0			
2			
4			
6			
8			
10			
12			
14			
16			
18			
20			
22			
24			
26			
28			
30			
32			
34			
36			
38			
40			
42			
44			
46			
48			
50			
52			
54			
56			
58			
60			

Laboratory #12
Some Area Properties of Cubic Curves:
An Exercise in Theorem Proving

Introduction

DERIVE is advertised as a 'Mathematical Assistant.' We have used it to handle difficult computations, do numerical searches, and provide us with graphic displays. As a result, we have solved difficult problems and made conjectures. In this section we will use DERIVE's symbolic capabilities to actually test a conjecture and prove some theorems about cubic curves.[4]

Before we begin our investigation, it is necessary to set up DERIVE to operate in a mode that is most natural for the type of calculations we will be doing throughout this lab. First begin by authoring $x^{\frac{1}{3}}$ and then plotting this expression. If your graph looks like the top one in Figure $12-1$, then you definitely need to change the setting using the Manage/Branch option. You will see that the word Principal is highlighted. Press "a" for Any. Now return and plot the expression again. You should see a graph similar to the one on the bottom of Figure $12-1$.

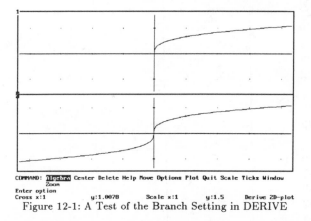

Figure 12-1: A Test of the Branch Setting in DERIVE

Even if your first graph looks like the one on the bottom of Figure $12-1$, you may still need to make some adjustments. Run one more test. Author the following expression

$$\frac{p^{\frac{1}{3}}}{(-p)^{\frac{1}{3}} + 2p^{\frac{1}{3}}}$$

---

[4]This project was originally designed by the Symbolic Computation Group at the University of Waterloo, Canada to be run on the Maple Computer Algebra System.

page 86

Simplify this expression. If the simplified result is 1, you are in the proper mode. If the result is the same as the above expression, you must use the Manage/Branch option. You will see that the word Real is highlighted. Press "a" for Any and and return to the main menu. Now simplify the expression, and you will have the desired result of "1."

The exact explanation for the above adjustments are best explained in a course on complex variables. All that will be said here is that the designers of DERIVE appreciated the fact that mathematicians work with number systems other than the real number system, and they were careful not to force their users into any one mode of operation.

Now that you have made the above adjustments, we are ready to begin our investigations of cubic curves. The majority of these investigations concern the area between a tangent line to the curve and the curve itself. We begin with a method for constructing a tangent line to a curve for a given value of the independent variable, say x.

Constructing a General Tangent Line Function

Recall that if $y = f(x)$, then the tangent line to the graph of y at $x = a$ is given by the expression

$$f(a) + f'(a)*(x - a).$$

The problem with entering this expression into DERIVE is one of evaluating both the function and its derivative at $x = a$. If we assume that $\lim_{x \to a} f'(x)$ exists, then we can write an expression equivalent to the above as follows:

$$\lim_{x \to a} y + \left(\lim_{x \to a} \frac{dy}{dx} \right)*(x - a).$$

Where $y = f(x)$ is differentiable at $x = a$. Note that we did not have to explicitly state that $\lim_{x \to a} f(x)$ exists, since that is covered by our differentiability hypothesis. We are now prepared to define a tangent line function in our DERIVE workspace. It is a little lengthy to type, but it will serve us throughout the entire lab.

$$\text{Tangent\_line}(y,x,a) := \lim(y,x,a) + \lim(\text{dif}(y,x),x,a)*(x - a)$$

Test your function by authoring x^3 and Tangent_line(x^3, x, -1). Simplify the latter expression and plot x^3 and the simplified expression. Now author Tangent_line(x^3,x,2), simplify and plot.

As an exercise to test your understanding of the process, construct a function called Secant_line (y, x, a, b) that gives the equation of a line passing through the graph of $y = f(x)$ at $x = a$

and $x = b$. Begin by writing an expression for the slope of the line using the evaluation method described above.

An Area Property of $y = x^3$

We are going to prove the following theorem. Refer to Figure $12 - 2$ to help clarify the text of the theorem.

> *Let $P \neq (0,\ 0)$ be any point on the graph of the curve $y = x^3$. The tangent line to the graph at P intersects the graph at a point, say Q. Likewise, if we draw a tangent to the curve at Q, it will intersect the curve at a point R. These tangent lines will lie on opposite sides of the curve. Furthermore, the area lying between the curve and the line segment QR is 16 times the area lying between the curve and the line segment PQ.*

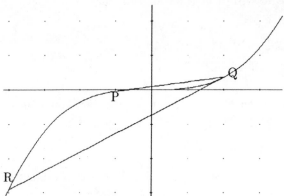

Figure 12-2: Illustating an Area Property of Cubic Curves

We have three things to prove in the above theorem

1. The tangent line to the graph of $y = x^3$ intersects the graph in at least one additional point.

2. The two tangent lines in question lie on opposite sides of the graph of $y = x^3$.

3. The second area is 16 times the first.

DERIVE can help us with all three of these proofs. We supply the brain power, DERIVE supplies the computational power.

Let's take on the first objective. Let p be the x-coordinate of the point, P. If we can show that there is another point that also lies on the line and the curve, we have accomplished the objective and proven #1. Author the following

$$\text{Tangent\_line}(x^3,\ x,\ p)\ =\ x^3$$

Simplify the expression and solve for x. Two new expressions will appear on the screen, $x = p$ and

$x = -2p$. This means that the point $(-2p, -8p^3)$ lies on both the graph of the tangent line and the curve $y = x^3$. Since this point is different from the point (p, p^3) for $x \neq 0$, the first part of the theorem is proved. Let Q be the point $(-2p, -8p^3)$. Note that the curve has different concavities at P and Q, since the two points are on opposite sides of $(0, 0)$, the inflection point of the curve. Refer to Figure $12 - 2$ to check out this fact. Thus, we have essentially proved #2 in the course of proving #1.

Continuing, solve the equation

$$\text{Tangent\_line}(x^3, x, -2p) = x^3$$

to find the point, R. The solutions for x will be $x = -2p$ and $x = 4p$. We now have the x-coordinates of the points of intersection of the lines and the curve, $y = x^3$. Evaluate the two integrals after simplifying the expressions for the integrants

$$\int_p^{-2p} (\text{Tangent\_line}(x^3, x, p) - x^3)\, dx \qquad \int_{-2p}^{4p} (x^3 - \text{Tangent\_line}(x^3, x, -2p))\, dx$$

These integrals are set up assuming the situation in Figure $12 - 2$. However, if the point P were chosen in the first quadrant, instead of the fourth, both integrals would be the negative of the respective areas. Since we are looking for the ratio of the areas, this will not affect our result.

Reference the expression numbers for the values of the above integrals and divide the value of the second integral by the value of the first. The value of the result is 16. We made no assumptions on our choice of a starting point, P, other than the one that P was not the inflection point of the curve, $(0, 0)$. Thus, the ratio of the areas is independent of P, and #3 has been proven.

The Laboratory Report

Since the graph of the curve $y = x^{\frac{1}{3}}$ is nothing more than the graph of $y = x^3$ reflected through the line $y = x$, a result similar to the one given above should hold. Refer to Figure $12 - 3$ and state this result for the graph of the cube root of x. After stating the result, use DERIVE to help you prove it. The reason we made the adjustments in the beginning of the lab was so that we could do the calculations for this curve.

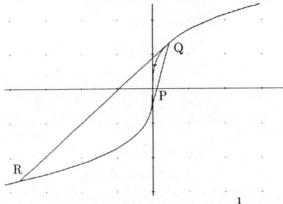

Figure 12-3: Extending the Result to $x^{\frac{1}{3}}$

A Second Result for Cubic Curves

Refer to Figure $12 - 4$ for the discussion that follows. Let $P \neq (0, 0)$ be any point on the curve and draw the tangent line to the curve at P. Let Q be the point of intersection of this line and the curve. From R draw a secant line to the inflection point at (0, 0). There are now two wedge-shaped areas. Find the area of each of these wedges. Show that the ratio of these two areas is a constant, independent of the choice of P. *Hint*: You already know the area of the large, undivided region, from doing the first part of the lab. Find the area of the lower region using the Secant_line function you developed earlier. The area of the upper region is the area of the large region minus the area of the lower region. Simplify your results prior to each calculation.

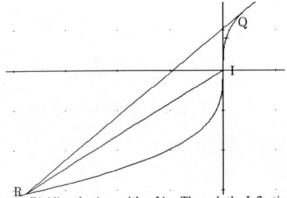

Figure 12-4: Dividing the Area with a Line Through the Inflection Point

Work Sheet

Laboratory #12

1. Starting at the point $(p, p^{\frac{1}{3}})$ find the other point of intersection of the tangent line to the graph of $x^{\frac{1}{3}}$ and the graph of the the function itself.

2. Using the result from problem 1 find the other point of intersection from this new point and the graph.

3. Write the expressions for the two integrals that give each of the areas of the two regions defined by the graph and the tangent lines.

4. Evaluate the ratio of the two integrals to prove the theorem for $f(x) = x^{\frac{1}{3}}$.

Laboratory #13

Finding a "Natural" Base:
An Exercise in Some Approximation Methods

Introduction

Exponential functions occur in many applications of mathematics. For example, if we have a population that doubles in size every 50 years, and its present size is 100,000, we know that in 150 years the size of the population will be 800,000 or $100,000*2^3$. In 75 years the population size will be about 282,800 or $100,000*2^{1.5}$. In fact, if x is measured in units of 50 years, then the population in x fifty year units from now will be $100,000*2^x$. Some other exponential functions of importance are $(1+\frac{r}{100})^x$, used in computing continuously compounded bank interest, $(\frac{1}{2})^x$, for studying radioactive decay; and 10^x as a representation of the real number system.

If we graph the function $f(x) = 2^x$ (look at the upper curve in the top display of Figure $13-1$), we see that it appears to be a continuous, smooth, concave-upward graph passing through the point (1,0). Authoring the function and using the Calculus/Differentiate operator followed by the approXimate operation shows us that the derivative of 2^x is approximately $.693147*2^x$. Thus, the derivative of this function is merely a constant times the function itself. At first this appears to be strange. However, if we consider the definition of the derivative and the properties of the exponential, we see that the result is to be expected

$$(2^x)' = \lim_{h\to 0}\frac{2^{x+h}-2^x}{h} = \lim_{h\to 0}\frac{(2^h-1)\,2^x}{h} = \lim_{h\to 0}\left(\frac{2^h-1}{h}\right)2^x = \left(\lim_{h\to 0}\left(\frac{2^h-1}{h}\right)\right)2^x$$

Thus, the constant multiplier of 2^x in the derived function is nothing more than the value of the limit

$$\lim_{h\to 0}\left(\frac{2^h-1}{h}\right).$$

The graph of 2^x and its derived function are shown in the top graph of Figure $13-1$. Note that the graph of the derived function always lies below the graph of 2^x.

The bottom graph in Figure $13-1$ is the function, $f(x) = 3^x$, and its derived function. Note that in this case, the graph of the derived function lies above the graph of 3^x. If you author 3^x in DERIVE, apply the Differentiate operation, and approximate the expression for the derivative, you will see that the derived function of 3^x is $1.09861*3^x$. Analysis similar to that done for 2^x

will show that the constant multiplier is the value of the limit

$$\lim_{h \to 0} \left(\frac{3^h - 1}{h} \right).$$

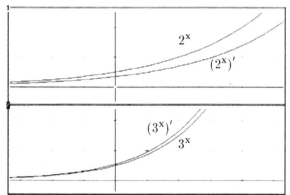

Figure 13-1: Graphs of Two Exponential Functions and Their Derived Functions

This raises an interesting question. Is there a base, b, for an exponential function such that

$$\lim_{h \to 0} \left(\frac{b^h - 1}{h} \right) = 1?$$

We can surmise that if there is such a base, it lies between 2 and 3. Furthermore, we might guess that lies closer to 3 than to 2. But, is the existence and value of such a b anything other than a mathematical curiosity? Let's think about this question.

Why Is a Function That Is Its Own Derived Function Important?

Let's suppose that we have a function, f(x), with the following properties

1. f(x) is a one-to-one function. This means that f^{-1} exists.
2. f is a differentiable function
3. f'(x) = f(x)

We will show the effect of these conditions on $(f^{-1})'$. Note that from the evidence submitted by the graphs of 2^x and 3^x, the function for which we are searching has these properties.

Recalling the Chain Rule and the definition of an inverse function,

$$f\!\left(f^{-1}(x) \right) = x \qquad \text{for all } x \in \text{Range of } f^{-1}.$$

$$\left(f\!\left(f^{-1}(x) \right) \right)' = 1$$

$$f'\!\left(f^{-1}(x) \right) \left(f^{-1}(x) \right)' = 1$$

But since $f'(x) = f(x)$, we see that the first derivative on the left-hand side of the above equation is merely $f(f^{-1}(x))$ or x. Therefore,

$$x \left(f^{-1}(x) \right)' = 1$$

and we find that $\left(f^{-1}(x) \right)' = \frac{1}{x}$.

In our calculus course we have seen that the function

$$\ln(x) = \int_0^x \frac{1}{t}\, dt$$

has a derivative (as a result of its definition) of $\frac{1}{x}$. Thus, our desired exponential function is the inverse function of the natural logarithm function, and the b we are searching for is the base for the natural logarithms.

Because of the great importance of the natural logarithms, our search for the base is more than a mathematical curiosity. It becomes a matter of computational necessity.

A Binary Search Method for Finding the Base

We begin by defining the following function in DERIVE.

$$f(b) := \lim_{h \to 0} \left(\frac{b^h - 1}{h} \right)$$

This is a rather strange looking function, but we have already evaluated it for $b = 2$ and $b = 3$. Clear the graphics window and center the plot so that it shows only positive values of the independent variable; then plot this function. This may take a while, depending on the model of computer you are using. When the graph is finally drawn, you will see that it appears to be a continuous, smooth, concave-downward curve that is defined for $x > 0$, i.e., it appears that f is a differentiable function of b for $x > 0$. At the moment all we will assume is the continuity of f.

Set the y-coordinate of the cross hair at 1 and move it horizontally until it touches the curve. You may obtain different values, but in a recent experiment, I obtained the result that the x-coordinate of the cross hair when it touched the curve with y-coordinate 1 was between 2.7083 and 2.7222. Thus, we estimate that b lies between 2.7083 and 2.7222. Approximating the value of f(b) at these points, we find that $f(2.7083) < 1 < f(2.7222)$. Thus, from our continuity assumption, we conclude that the value of b we want lies between these two values that are separated by .0139 . Thus we know the value of the base to one decimal place: it is 2.7. Our strategy will be to choose a point between these two values (2.71525) and evaluate the function at this approximation of b. If the function value is

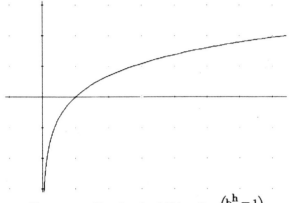

Figure 13-2: The Graph of $f(b) = \lim_{h \to 0} \left(\frac{b^h - 1}{h} \right)$

less than 1, we know that our desired value is between 2.71525 and 2.7222 . If the function value is greater than 1, we know that the desired value is between 2.7083 and 2.71525. Whatever the case, we halve the interval and repeat the process in the appropriate interval. We continue this process, choosing the appropriate interval, until we have obtained the desired accuracy for a value of the base.

DERIVE can assist us in this process. We begin by entering 2.71525 and f(b) on separate lines, say m and m+1. Now author f(#m) and approximate the value. It will be less than 1. Now author 2.7222 on line, say n, and author (#m+#n)/2. Approximate this sum and evaluate f for this value. Continue choosing the midpoint of an interval for which

$$f(\text{left-hand endpoint}) < 1 < f(\text{right-hand endpoint})$$

The desired value of the base will always lie between the left and right endpoints. After 10 iterations of this procedure you should have an approximation to the base that is accurate to four decimal places, i.e., you should know that the number is 2.7182 . . . . Record each step of the above procedure for 10 steps.

This procedure is, in general, rather slow, and if we can assume the differentiability of the function, it can be replaced with a much faster procedure that is attributed to the great mathematician and physicist, Isaac Newton, one of the inventors of the calculus.

Newton's Method

Assume that F(x) is differentiable on an interval (a, b) and that F(x) = 0 for some x in the interval. Such is the case for f(x) − 1 in the interval (0, 6) shown in Figure 13 − 3, where f(x) is the function defined on the previous page.

page 95

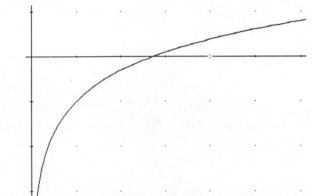

Figure 13-3: A Function Satisfying the Hypotheses for Newton's Method

We draw a tangent line to the curve at one of the end points, 2.70083 in the case of Figure $13 - 3$, and find the point in the interval where the tangent intersects the x-axis. This will be our new approximation of the value, and we draw a tangent to the curve at this point. We repeat the process until the values of the point of intersection of the tangent line and the x-axis have reached a desired accuracy. The last value of x obtained by this process is the approximation of the root of the function. A graph of this process is given in Figure $13 - 4$.

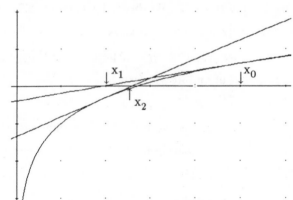

Figure 13-4: Two Iterations of Newton's Method to Find the Root of $f(x) - 1$

An analytic description of this process follows. Assume that x_i is the value of the approximation after i iterations. We will assume that $x_0 = a$. Recall that the equation of a line passing through a point, (α, β) with slope, m, is

$$y - \beta = m*(x - \alpha).$$

In our case, $\alpha = x_n$, $\beta = f(x_n)$, and $m = f'(x_n)$. Furthermore, we want to find the value of x when $y = 0$. Thus, the above equation becomes

$$0 - f(x_n) = f'(x_n)*(x_{n+1} - x_n)$$

Solving for x_{n+1}, we have

$$x_{n+1} = x_n - \frac{f(x_n)}{f'(x_n)}$$

This gives us an easy way to find the next approximation. We merely subtract from the previous approximation the value of the function at the approximation divided by the value of its derived function at the approximation.

Using the DERIVE trick of taking limits to evaluate functions and derivatives at specific points, we can author the following function to be used in Newton's method.

$$\text{Newton}(u, x, a) := a - \lim(u/\text{dif}(u, x), x, a)$$

Recall that this usage makes an additional assumption on the continuity of the derived function at the point a. The picture of our graph of $f(b)$ gives us confidence that this assumption is valid for $f(b)$.

We are now ready to apply Newton's method to $f(b)$. We begin by authoring

$$\text{Newton}(f(b) - 1, b, x)$$

Use the Options/Precision sequence to set the precision to 20. Begin by using Manage/Substitute to substitute 2.7083 for x. Approximate this value. Note the line number of the approximation. Using this line number, substitute the new value into the Newton expression for x and approximate it. Continue until the approximation stabilizes, i.e., does not change its value except in the very last decimal place. This happens after about four iterations. The approximation to the value of the base is

$$2.71828188284590452353.$$

If we compare this to the value that DERIVE has stored for $\hat{e}$, the name given to the base of the natural logarithms, you will see that it agrees in every decimal place. As a further check, evaluate $f(b)$ for this approximation, and you will get a value of 1. We accept the above number as an approximation of the base we were seeking, based on this evidence.

The Laboratory Report

The above procedure made some assumptions about the behavior of the exponential function and the function, $f(b)$. These assumptions can be avoided by taking another approach to the problem. If we change our view to one of looking for the base of the natural logarithm function, $\ln(x)$, we do not need to make any guesses or assumptions about the differentiability of the function and the continuity of the derivative. We know that $(\ln(x))'$ is $\frac{1}{x}$, and as long as we stick with positive x, we have no trouble meeting the assumptions implied by our methods.

Since we are looking for the base of $\ln(x)$, we are looking for that number, x, for which $\ln(x) = 1$. Thus, we can (safely) apply Newton's method to $\ln(x) - 1$. Start with $x = 2.7$ and apply Newton's method to $\ln(x) - 1$. How many iterations did the approximation method take using this function and starting point? Report on your method and results in a manner that is understandable to a student in a calculus class at another school that has not done this lab.

DERIVE 2.0

Using the Iterates Procedure for Newton's Method

The Iterates procedure lends itself very nicely to Newton's root-finding method. One can implement the method by defining a function $Newton(u, x, s, n)$, where

$$x = \text{the independent variable}$$
$$u = \text{the expression for which a root is desired}$$
$$s = \text{the initial "guess"}$$
$$n = \text{the number of iterations}$$

The function is defined as follows:[5]

$$Newton(u, x, s, n) := iterates(x - u/dif(u,x), x, s, n).$$

The root of $\ln(x) - 1$ located near $x = 2$ can be found by simplifying the function

$$Newton(\ln(x) - 1, x, 2, 10).$$

---

[5]The definition used here for the Newton function is taken from the DERIVE instruction manual that is supplied with the DERIVE package, version 2.0.

Work Sheet

Laboratory #13

1. Graph the function ln(x). Set the cross hair on y = 1 and move it left or right to determine two
   values of x such that the graph of ln(x) lies between these two values. Using these values as the
   initial iteration, apply the binary search routine to estimate the value of x for which ln(x) = 1.

Lower estimate	Upper estimate	Median

2. Author a Newton's Method function. Start with x = 2.7 and apply the method to the function
   ln(x) − 1 for five iterations.

Iteration #	Value of x
0	2.7
1	
2	
3	
4	
5	

Laboratory #14

**Antiderivative Formulas Involving
Exponential and Logarithmic Functions**

Introduction

Pattern matching and pattern recognition are skills that we ordinarily associate with disciplines such as computer science and the natural sciences.   They are also valuable skills for mathematicians as well as those who use mathematics.   Many important breakthroughs in mathematics have centered about the realization that the behavior of one system is similar to a behavior in another, seemingly unrelated system.   In this lab you will be looking for patterns of behavior to classify general formulas for families of integrals of the form

$$\int x^n f(x)\,dx \qquad \text{and} \qquad \int \big(f(x)\big)^n dx$$

where $f(x)$ is an exponential, logarithmic function or the derivative of such a function. We begin by looking at the integral analog of the product rule and using DERIVE to implement our calculations.

The product rule, using differential notation, states that the product of differentiable functions $u(x)$ and $v(x)$ has a differential

$$d(uv) = u\cdot dv + v\cdot du$$

Rearranging this equation shows that

$$u\cdot dv = d(uv) - v\cdot du$$

or, taking the antiderivative of both sides of the above equation

$$\int u\cdot dv = uv - \int v\cdot du + C$$

for some constant, C.   This formula is known to students of calculus as the "integration-by-parts formula."   Sometimes it gives us a very handy way to transform antiderivatives that we do not recognize into a form that we can evaluate. For example, consider the expression

$$\int x\,e^x\,dx$$

If we let $u = x$ and $dv = e^x dx$, then it is possible to transform the integral into a form we recognize.

$$\int x\,e^x\,dx \;=\; x\,e^x \;-\; \int e^x\,dx \;+\; C$$

$$=\; x\,e^x \;-\; e^x \;+\; C$$

Check this result by substituting the appropriate values for u and v in the integration-by-parts formula. Another result we can obtain by letting $u = \ln(x)$ and $dv = dx$ is

$$\int \ln(x)\,dx \;=\; x\cdot\ln(x) \;-\; x \;+\; C.$$

This result is easily checked when you note that $v = x$ and $v\cdot du = \frac{x}{x}\,dx = dx$. After you have checked these integrals by hand, test DERIVE's ability by having it evaluate the two integrals given above.

Finding a Formula for the Antiderivatives of a Family of Functions

We will use DERIVE to help us discover the formula for the following family of antiderivatives

$$\int \frac{x^n}{x+b}\,dx \qquad n=0,1,2,\ldots$$

Since $\int \dfrac{1}{x+b}\,dx \;=\; \ln(x+b) \;+\; C$, we have the value of the integral for $n = 0$, i.e.,

$$\int \frac{x^0}{x+b}\,dx \;=\; \int \frac{1}{x+b}\,dx \;=\; \ln(x+b) \;+\; C$$

This is the cornerstone of our investigation, and DERIVE is the tool we will use to generate our data. Use DERIVE to author the following expression for the general term of the family of antiderivatives.

$$\int \frac{x^n}{x+b}\,dx$$

Note that if we try to simplify this expression, DERIVE merely returns the expression. Since it has no way of knowing that n is a non-negative integer, it is not possible to evaluate the antiderivative. Thus, we cannot take the easy way out and simply have DERIVE do all of the work for us. We will use the Manage/Substitute option to substitute consecutive values for n and examine the relationship between the different values of the antiderivative. Recall that DERIVE always sets the constant $C = 0$. Thus, the antiderivatives given in this text are specific functions with C chosen to be 0. Let's just evaluate the integral for $n = 1, 2, 3$, and 4. We will then look for a pattern and test it for $n = 5$.

$n = 1$

$$\int \frac{x}{x+b}\,dx \;=\; x \;-\; b\cdot\ln(x+b)$$

n = 2

$$\int \frac{x^2}{x+b} \, dx = \frac{x^2}{2} - b \cdot x + b^2 \ln(x+b)$$

n = 3

$$\int \frac{x^3}{x+b} \, dx = \frac{x^3}{3} - b\frac{x^2}{2} + b^2 x - b^3 \ln(x+b)$$

n = 4

$$\int \frac{x^4}{x+b} \, dx = \frac{x^4}{4} - b\frac{x^3}{3} + b^2\frac{x^2}{2} - b^3 x + b^4 \ln(x+b)$$

Looking at the above, we observe several patterns.

1. Every antiderivative has a term involving $\ln(x+b)$.

2. The coefficient of $\ln(x+b)$ is $\pm b^n$. The $+$ sign is in the even terms, the $-$ sign is in the odd terms.

3. The remaining terms of the antiderivative are a polynomial in b and x and the exponents of b and x always add to n. (Recall, $b = b^1$. Likewise for x.)

4. The x^i term of the polynomial is always over i for $i = 1, 2, \ldots, n$.

5. The signs of the terms of the polynomial alternate between $+$ and $-$.

6. The sign of the $b^{n-1}x$ term is always the opposite of the $b^n \ln(x+b)$ term.

A favorite gimmick of mathematicians is to write alternating signs as powers of -1. This is because even powers of -1 are equal to 1, and odd powers are equal to -1. Thus, $(-1)^n$ alternates signs as n increases. Combining observations 1 and 2 above gives us

$$\int \frac{x^n}{x+b} \, dx = (-1)^n \, b^n \ln(x+b) + \text{a polynomial in b and x}$$

$$= (-b)^n \ln(x+b) + \text{a polynomial in b and x.}$$

Now, we look at the polynomial term. Observation number 3 leads us to the fact that every term of the polynomial is of the form

$$\text{some coefficient} * b^{n-i}x^i$$

Observations 4, 5, and 6 help us to construct the coefficient. For the x^i term, it is of the form

$$\frac{(-1)^{n-i}}{i}$$

Thus, the general term of the polynomial is

$$\frac{(-b)^{n-i}x^i}{i} \qquad \text{for } i = 1, 2, \ldots, n-1.$$

To get the complete polynomial, we merely add these terms, using summation notation. Thus the result for the antiderivative is

$$\int \frac{x^n}{x+b}\,dx = (-1)^n b^n \ln(x+b) + \sum_{i=1}^{n-1} \frac{(-b)^{n-i} x^i}{i}$$

Note that this formula even holds when n = 0.

Use this result and the 6 observations to find the value for

$$\int \frac{x^5}{x+b}\,dx$$

Check your result with DERIVE. If you are still are not comfortable with this formula, try developing formulas when n = 6, 7, . . .

A formula such as we developed at the bottom of the previous page is called a *summation formula*. It gives us a rule for directly developing the value of the antiderivative for each n. We merely plug in the value of n and evaluate the sum. Such formulas are useful and usually very efficient tools for the task at hand. We will develop at another way to determine the value of the integral by looking at its value in terms of the values of the previous integrals. Such formulas are called *recurrence formulas*.

Recurrence Integral Formulas

If you were observing the process of evaluating the integral, you may have made the following observation

Each successive integral is of the form $\frac{x^n}{n} - b\cdot$ *the previous integral, or*

$$\frac{x^n}{n} - b\cdot \int \frac{x^{n-1}}{x+b}\,dx .$$

The validity of this statement can be shown rather easily by differentiating the above expression and setting it equal to $\frac{x^n}{x+b}$. The trouble with a recursive formula is that you have to keep applying it until you get to a case that you know. For example, when n = 4, we have

$$\int \frac{x^4}{x+b}\,dx = \frac{x^4}{4} - b\cdot \int \frac{x^3}{x+b}\,dx = \frac{x^4}{4} - \frac{bx^3}{3} + b^2\cdot \int \frac{x^2}{x+b}\,dx$$

$$= \frac{x^4}{4} - \frac{bx^3}{3} + \frac{b^2x^2}{2} - b^3\cdot \int \frac{x}{x+b}\,dx$$

$$= \frac{x^4}{4} - \frac{bx^3}{3} + \frac{b^2x^2}{2} - b^3\cdot x + \int \frac{1}{x+b}\,dx$$

$$= \frac{x^4}{4} - \frac{bx^3}{3} + \frac{b^2x^2}{2} - b^3\cdot x + b^4\cdot \ln(x+b).$$

Eventually we reduced the antiderivative to the base case, or $\int \frac{1}{x+b}\,dx$, an integral we know. When we reach that point, we can evaluate the integral and stop the evaluation process. Without this base case, we would be unable to evaluate the integral. Generally speaking, a recurrence formula is made up of two parts: the general case, that tells us how the value at step n is obtained from the value at step $n-1$, and a base case, that tells us how to stop the process. The recurrence statement of the formula is more compact to state, but requires n applications of the formula to evaluate the integral. On the other hand, the summation formula requires a more detailed statement but is, in general, more efficient to evaluate for one particular case. Tables of Integrals have examples of both types of formulas.

The Laboratory Report

In this report you will use DERIVE to generate a general formula for the following integrals for $n = 1, 2, 3, 4, \ldots$.

1. $\int (\ln x)^n\,dx$

2. $\int x^n \ln(x+b)\,dx$

3. $\int x^n e^x\,dx$

For the first integra,l find both a summation and a recurrence formula. For the second integral, find the summation formula only. For the third integral, find a recurrence formula only. In the case of the second integral, Simplify does not expand the non-logarithmic term enough to recognize the complete pattern. The answer is in two terms. The form of the logarithmic term is easy to spot, but the form of the other term, as it stands, is difficult to recognize. Highlight this term for each n and press "e" for Expand. This will expand the term and make a general pattern more easy to see.

In each case note your observations about the integrals and discuss how you built up the general formula. Check your result by differentiating and simplifying the resulting derivative to show that the derivative is the integrand.

Name_____

Course_____

Work Sheet

Laboratory #14

Fill in the table for each value of n.

n	$\int (\ln x)^n \, dx$	$\int x^n \ln(x+b) \, dx$	$\int x^n e^x \, dx$
1			
2			
3			
4			
5			

Use the above data to make a conjecture about the general form of the following integrals

$$\int (\ln x)^n \, dx \ = $$

$$\int x^n \ln(x+b) \, dx \ = $$

$$\int x^n e^x \, dx \ = $$

Laboratory #15

Applications of the Exponential Function:
Some Models of Population Growth

Introduction

The exponential function is unique among the class of all differentiable functions in that it is a function that is its own derivative. Thus, if $y = Ce^x$ for some constant, C, then y satisfies the equation

$$y' - y = 0 \qquad \text{or} \qquad y' = y.$$

Equations such as these that involve a function and its derivatives are called *differential equations*. A differential equation, such as the above, that involves only the first derivative of a function and possibly the function and the independent variable is called a *first order differential equation*. During part of our investigation we will be considering a special class of first order differential equations, called *linear first order differential equations*. These equations have the form

$$y' + p(x) \cdot y = q(x)$$

where p and q are known functions of the independent variable, x. The equation satisfied by $y = Ce^x$ is a special case of a linear equation with $p(x) = -1$ and $q(x) = 0$. We will see that the exponential function plays an important role in solving all first order linear differential equations.

The growth of populations is characterized by the fact that in the absence of any constraining forces, the rate of change in the size of the population is directly proportional to the present size of the population. For example, if P(t) is the size of the population at time, t, then

$$\frac{dP}{dt} = rP.$$

If we let P' denote the first derivative of P with respect to t, then we have the first order linear differential equation

$$P' - rP = 0.$$

Before continuing our discussion of population growth, we will investigate the solution of the general first order linear differential equation and the role of the exponential function in this solution. We will also implement the general solution method in DERIVE. After these mathematical preliminaries, we will continue discussing models for population growth. We will use DERIVE and our general solution to examine the solutions resulting from the models.

Solving First Order Linear Differential Equations

Suppose that $u = u(x)$ is a differentiable function of x. Using the differentiation property of the exponential function and the Chain Rule, we see that

$$\frac{d(e^u)}{dx} = e^u \cdot \frac{du}{dx} = e^u \cdot u'.$$

This fact may seem unrelated to the primary goal of solving a first order linear differential equation, but it will prove to be one of the essential steps in the solution process. Another important observation regards the operation of the product rule for derivatives. In particular, if u and v are both differentiable functions of x, then

$$(u \cdot v)' = u \cdot v' + v \cdot u'.$$

Turning now to the general first order linear differential equation,

$$y' + p(x)y = q(x),$$

we see that the left hand side bears some similarity to the derivative of a product. More properly, y behaves as if it is being differentiated in conjunction with a product. The question is: what is the other term in the product? The answer to this question is supplied by our knowledge of the exponential function and a shrewd observation.

Suppose that $\phi(x)$ is a differentiable function with the property that $\phi'(x) = p(x)$, i.e., $\phi(x)$ is an antiderivative of p(x), or $\phi(x) = \int^x p(x)\,dx$. Given this condition, then

$$(e^{\phi(x)}y)' = e^{\phi(x)}y' + (e^{\phi(x)})'y = e^{\phi(x)}y' + e^{\phi(x)}p(x)y$$

$$= e^{\phi(x)}(y' + p(x)y).$$

Thus, if we multiply both sides of the differential equation by $e^{\phi(x)}$, we have

$$e^{\phi(x)}(y' + p(x)y) = q(x)e^{\phi(x)}.$$

From the above, this is equivalent to

$$(e^{\phi(x)}y)' = q(x)e^{\phi(x)}.$$

Integrating both sides of the equation with respect to the independent variable, x, we have

$$e^{\phi(x)}y = \int q(x)e^{\phi(x)}\,dx + C.$$

From this, we easily move to the solution

$$y = e^{-\phi(x)}\left(\int q(x)e^{\phi(x)}\,dx + C \right).$$

Recall that $\phi(x)$ is an antiderivative of p(x). This means that the expression $e^{\phi(x)}$ is actually $e^{\int p(x)\,dx}$. This expression is called an *integrating factor* because when we multiply both sides of the equation by this factor, the left side becomes the derivative of a product. Hopefully, the product on the right side, $q(x)\,e^{\phi(x)}$, will be a function that we can integrate. In the case of the equations we have mentioned thus far, $q(x) = 0$, and the product on the right is trivial. This is not always the case, but more on this later. Let's implement this solution in DERIVE.

Each of DERIVE's operators can be written as part of an expression as well as activated from the menu by key strokes. For example, the operation of integrating p(x) with respect to x is written as "int(p(x),x)" and appears on the screen as $\int p(x)\ dx$. Allowing u to denote the function p(x) and v to denote q(x) in the general first order linear differential equation, we author the following function (remember that ê is entered via the Alt- e combination)

$$\text{lin\_de1}(p,q,,x) := \hat{e}\,\hat{\ } - \text{int}(p,x)*(\text{int}(q*\hat{e} \wedge \text{int}(p,x),x)\ +\ c)$$

The resulting screen display is

$$\text{LIN\_DE1}(p,q,x) := \hat{e}^{\,-\int p\ dx}\left(\int q\ \hat{e}^{\int p\ dx}\ dx\ +\ c\right).$$

If a first order linear differential equation has a solution, this function will enable us to find the solution. This is done by calling the function with the expression for p(x) substituted for p and the expression for q(x) substituted for q.

Test the function by authoring LIN_DE1(-1, 0, x) and simplifying. The result should be $c\hat{e}^{x}$. Now try LIN_DE1($-r$, 0, x). Substitute .02 (representing a 2% growth rate) for r and 1 for c. Draw the graph of this function using an x-scale of 20 and a y-scale of 5 (see Figure $15-1$).

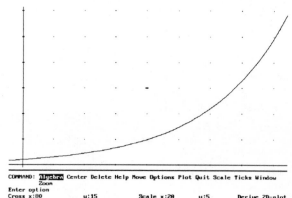

Figure 15-1: The Solution to the Linear Differential Equation y' - .02y = 0 with y(0)=1

Linear First Order Differential Equations as Models of Population Growth

In the introduction to this lab we say that if a population has a growth rate, r, then it satisfies the differential equation

$$P' - rP = 0.$$

At the end of the last section we substituted .02 for r and graphed the resulting function. It is clear that this function will exceed any bound that we care to place upon it. The alarming fact about this function is that the larger the function becomes, the faster it grows. If a population were to grow according to this equation, it would eventually outgrow the carrying capacity of its environment. This is exactly the point made by the British economist, Thomas Malthus (1766 − 1834) who predicted war, famine, and plague as the ultimate future for mankind. Indeed, Malthus' predictions would be correct if no controls were placed upon the growth rate of the population.

Suppose that there are other forces acting on the rate of population growth in addition to the pressure from the population size. In particular, let's assume that there is a force that is proportional to t, the independent variable. Such a force can be justified by the growth of medical technology that prolongs life expectancy and agricultural technology that increases the available food supply. Such a force can be represented in the differential equation in the following way

$$P' = rP + \alpha t$$

Note that a positive value for α increases the growth rate, and a negative value has a retarding effect. If we place the differential equation in its standard form,

$$P' - rP = \alpha t ,$$

we can use DERIVE to solve the equation by authoring Lin_del(-r, αt, t) and simplifying this expression. Note: α is authored as Alt-a. Within a few seconds, DERIVE returns the following result:

$$P(t) = C e^{rt} - \frac{\alpha(rt + 1)}{r^2}$$

The following table gives the U.S. Census Bureau figures for the decades from 1950 until 1990.

Year	U.S. population (in millions)
1950	150.697
1960	178.464
1970	204.766
1980	226.5
1990	247.1 (projected)

In order to facilitate the solution for the values of C and α based on these data, we will assume that the growth rate based solely on the size of the population, P, is 1.5%, or r = .015. Using the 1950 and 1970 values of the population, where t=0 in 1950, we have a system of two equations in two unknowns that can be solved to yield C = 177.773 and α = .00607563. These solutions were found using the Manage/Substitute option in DERIVE to construct the equations and then solving the resulting system of equations. Figure $15-2$ shows the graph of P(t) after substituting r = .015, α = .00608, and C = 177.773.

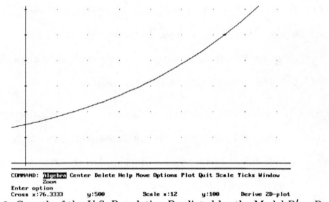

Figure 15-2: Growth of the U.S. Population Predicted by the Model $P' - rP = \alpha t$

Note that in this figure the cross hair lies on the curve at t = 76.333, P = 500. If we assume that the carrying capacity of the United States is 500 million, this model predicts that we will exceed this capacity in the year 2026 (1950 + 76), a rather ominous prediction.

There are several interesting experiments that you can do with this model and extensions to the model. For example, the choice of r = .015 was certainly arbitrary. How does the model behave for other values of r? What happens when α is negative? How well does the model fit the existing data? Suppose the additional force governing the population growth is proportional to t^2 as well as to t? There are many more interesting questions that can be asked about this model and others arising from first order linear differential equations. DERIVE supplies us with a tool to investigate them.

A Nonlinear Model for Population Growth

In 1840 the Belgian mathematician and demographer, Peter Verhulst, constructed a model for population growth based on the United States census data from the years 1790 to 1840. This model has a feature that makes it qualitatively different from those we have seen in the previous examples. Verhulst's model contains a term to account for competition between individual members of the population. In particular, Verhulst concluded that the effect of competition on population growth is directly proportional to the number of pairs of individuals within the population. Thus, if a population has P individuals, the number of competing pairs is given by the well-known combinatorial formula

$$\frac{1}{2} \cdot P \cdot (P-1)$$

Subtracting a multiple of this term from the right side of the basic model for population growth gives us Verhulst's model

$$P' = \alpha P - \beta P^2$$

Although this is a first order differential equation, it is no longer a linear equation, since we have P appearing to the second power. Consequently, we will have to develop a different technique to solve the equation.

We decide to begin by factoring α from the right-hand side. Letting $\delta = \frac{\beta}{\alpha}$, we have the equation

$$P' = \alpha (P - \delta P^2)$$

This is equivalent to

$$\frac{P'}{P - \delta P^2} = \alpha$$

Integrating both sides of this equation with respect to t we note that in differential notation,

$$dP = P' dt$$

we have the integral equation

$$\int \frac{dP}{P - \delta P^2} = \int \alpha \, dt + C \ .$$

After performing the integration, we have an equation in P and t that can be solved for the dependent variable, P, in terms of the independent variable, t. Before using DERIVE to solve this equation, we will look at a more general case inspired by this example.

For this discussion, assume that y is the dependent variable, and x is the independent variable. Furthermore, assume that u is some function of x alone, and v is a function of y. Consider the equation

$$y' = u \cdot v \ .$$

An equation of this form is called a *variables-separable differential equation*, since it is possible to isolate x and y on opposite sides of the equation as we will see very shortly. Following a procedure similar to our actions above, we divide both sides of the equation by v, the function of y, and note that $dy = y' dx$. Integrating both sides, we have the integral equation

$$\int \frac{dy}{v} = \int u \, dx + C \ .$$

Just as we summarized the process for solving linear differential equations as a DERIVE function, we can encapsulate the process for solving a separable differential equation in a DERIVE function definition. Author the following expression

$$\text{sep\_de1(u,v,x,y)} := \text{int}(1/v,y) = \text{int}(u,x) + c \ .$$

The screen display of the above is

$$\text{SEP\_DE1(u,v,x,y)} := \int \frac{dy}{v} = \int u \, dx + c \ .$$

This expression is much different than any previous examples we have done in the past. We are actually assigning the value of a function to be an equation. This is unusual, but perfectly legal in DERIVE and as a function definition.

We will now solve Verhulst's model of population growth. Author and simplify the following

$$\text{sep\_de1}(\alpha, \ P - \delta P^2, \ t, \ P) \ .$$

This results in the following equation (Note that DERIVE may translate P to p. Also, δ is authored as Alt-d)

$$\text{Ln}\left(\frac{p}{p\delta - 1} \right) = c + t\alpha \ .$$

Authoring and simplifying $\hat{e}\,\hat{} \#M$, where M is the DERIVE expression number of the above equation, we have the equivalent equation

$$\frac{p}{p\delta - 1} = \hat{e}^{\,c + t\alpha}$$

Recognizing that e^c is a constant, we author the equation

$$\frac{p}{p\delta - 1} = a\hat{e}^{t\alpha} \ .$$

Solving for p, we see

$$p = \frac{a\hat{e}^{t\alpha}}{a\delta\hat{e}^{t\alpha} - 1}$$

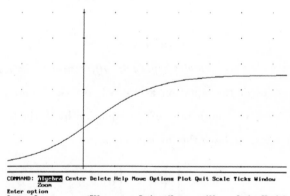

Figure 15-3: U.S. Population Growth Predicted by the Verhulst Model

Using the U. S. population data from the table on the bottom of page 109, the following values can be obtained $\alpha = .025$, $\beta = .00007$, $\delta = .0028$, and a $= -260.7$. The graph in Figure $15 - 3$ is the graph of p with these values for α, β, δ, and a having been substituted in the right side of the equation for p. Note that this curve definitely appears to approach a horizontal asymptote. This is indeed the case. If we refer to the original model, we see that $P' = 0$ when $P = \frac{\alpha}{\beta}$. In the above example, this means that $P' = 0$ when $P \approx 357$ million. This value is called the *equilibrium* value of the population. Careful examination of Figure $15 - 3$ shows that the values of P are approaching this equilibrium value. The Verhulst model, in effect, predicts that the U.S. population will not exceed this value. This may be an overly optimistic prediction; however, several tests of the model have shown that the results are quite accurate in the short run. On the other hand, some of the equilibria predicted by the model have been exceeded.

To see why the Verhulst model may not be appropriate for long-range predictions of population, let's look at what happens if the population size exceeds the equilibrium value. The graphs in Figure $15 - 4$ illustrate that in the long run the solution given by the model always tends to the equilibrium value. In the top graph the value of a was chosen so that $P = 500$ million when $t = 0$. Note that the population steadily decreases to asymptotically approach the equilibrium value of 357 million. This is certainly a very optimistic outlook on population dynamics. It is improbable that the equilibrium value can be that stable.

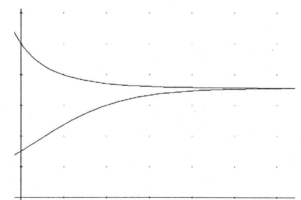

Figure 15-4: An Illustration of the Extreme Stability of Verhulst's Equilibrium Value

The Laboratory Report

Choose one of the situations given below and construct a differential equation model of the situation. Use one of the DERIVE procedures given in the narrative for the lab and solve the model. In your report discuss the construction of the model and include an analysis of your solution.

1. You enroll in an IRA plan that yields 6% interest compounded continuously. Each month, according to the terms of the plan, you deposit a fixed amount, A, in the account. What will be the value of the account when it reaches maturity in 20 years? You determine the value for A and allow your time unit to be one month.

2. You are a homicide detective and you must determine the time of death of a murder victim. You enter a room that is 70°F. The body temperature of the victim is 87°F, and one hour later the victim's body temperature is 82°F. Use Newton's law of cooling to determine the approximate time the victim died. Assume that the victim had a normal body temperature of 98.6°F.

3. The velocity of an object falling under the influence of gravity increases due to the force of gravity, $g = 9.8$ m/sec^2, and is retarded by air resistance that is proportional to the square of the velocity, i.e., the retarding force is of the form kv^2. Show that this retarding force imposes a limiting velocity on a freely falling object. If the limiting velocity for a freely falling human is 50 m/sec, how long does a parachutist jumping from a height of 1200 meters have until she reaches a height of 500 meters when the parachute must be opened?

4. A tank contains 1000 liters of distilled water, and a brine solution containing 0.3 kilograms of salt is being pumped into the tank at a rate of 2 liters per minute. Water is being withdrawn from the tank at the same rate. How long will it take until the tank contains a 2% brine solution?

Work Sheet

Laboratory #15

1. Apply the function LIN_DE1 developed in the manual to solve the following linear differential equations.

$$y' + 3y = x^2$$

$$y' - \frac{2}{x}y = x^3$$

$$y' = x - 3y$$

$$y' - my = a e^{mx} \qquad \text{(a and m constants)}$$

2. Apply the function SEP_DE1 to solve the following separable differential equations.

$$xyy' = 2(y+3)$$

$$(1-x)y' = y^2$$

$$my = nxy' \qquad \text{(m and n positive integers)}$$

$$\frac{dV}{dP} = -\frac{V}{P}$$

Laboratory #16

Finding the Value of Infinite Series

Introduction

Let $\{a_n\}$ be a sequence of real numbers. By definition, the value of an infinite series, $\sum_{i=1}^{\infty} a_i$, is given by

$$\sum_{i=1}^{\infty} a_i = \lim_{n\to\infty} \sum_{i=1}^{n} a_i = \lim_{n\to\infty}(a_1 + a_2 + \cdots + a_n).$$

Most of the results studied in calculus deal with the existence of the limit on the right side of the above equation. The subject we will address in this laboratory is the actual evaluation of the limit. This task is not as easy as it may first appear. For example, suppose $\{a_n\} = \left\{\frac{1}{n^2}\right\}$. Authoring $\frac{1}{i^2}$ and using the Calculus/Sum option generates the expression

$$\sum_{i=1}^{n} \frac{1}{i^2}$$

Applying the Calculus/Limit option to this expression yields

$$\lim_{n\to\infty} \sum_{i=1}^{n} \frac{1}{i^2}$$

This is exactly the expression we want. However, an attempt to simplify or approximate the expression only yields another expression

$$\sum_{i=1}^{\infty} \frac{1}{i^2}.$$

The result is a correct symbolic result, but not very helpful for determining the value of the series. This may seem to be a weakness in DERIVE, but it is unreasonable to expect DERIVE to have the symbolic routines to provide an exact evaluation for every possible series. If we return to the summation and replace the variable n by 500, we can estimate the value of the limit as 1.64293. But how accurate is this estimate? The fact that we have summed 500 terms of the sequence does not in itself guarantee an accurate estimate. After all, if we evaluate $\sum_{i=1}^{500} \frac{1}{i}$, the value given is 6.79282, and that is a long way from indicating that the series is unbounded. We will consider the question of finding the value of positive-term series as part of our next activity in this lab. We will use the methods employed in the proofs of the integral test and the ratio test as the basis for approximation methods for series whose individual terms form a positive, decreasing sequence.

The Integral Test for Positive Term Series

Recall the statement of the integral test for the convergence of an infinite series:

If $f(x)$ is a nonnegative decreasing function such that $f(n)=a_n$ for $n=1,2,\ldots,$

then the series $\sum\limits_{i=1}^{\infty} a_n$ converges if and only if $\int\limits_{1}^{\infty} f(x)\,dx$ converges.

The proof of this statement relies on the fact that if $f(x)$ is as described, then

$$\int\limits_{N+1}^{\infty} f(x)\,dx \ \leq\ \sum\limits_{i=N+1}^{\infty} a_i \ \leq\ \int\limits_{N}^{\infty} f(x)\,dx \ .$$

This fact is illustrated in Figure $16-1$. Note since our partition is 1 unit wide, the area of the ith rectangle is $f(i)\cdot 1$ or a_i .

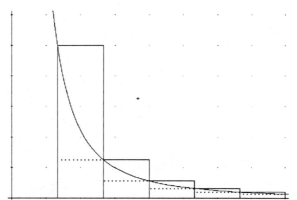

Figure 16-1: A Visual Proof of the Integral Test for Convergence of a Series

Returning to the basic inequality used in the proof of the statement, we have the following set of equivalent inequalities.

$$\int\limits_{N+1}^{\infty} f(x)\,dx \ \leq\ \sum\limits_{i=N+1}^{\infty} a_i \ \leq\ \int\limits_{N}^{\infty} f(x)\,dx$$

Subtracting $\int\limits_{N+1}^{\infty} f(x)\,dx$ from each term,

$$0 \ \leq\ \sum\limits_{i=N+1}^{\infty} a_i - \int\limits_{N+1}^{\infty} f(x)\,dx \ \leq\ \int\limits_{N}^{\infty} f(x)\,dx - \int\limits_{N+1}^{\infty} f(x)\,dx$$

$$0 \ \leq\ \sum\limits_{i=N+1}^{\infty} a_i - \int\limits_{N+1}^{\infty} f(x)\,dx \ \leq\ \int\limits_{N}^{N+1} f(x)\,dx$$

$$0 \ \leq\ \left| \sum\limits_{i=N+1}^{\infty} a_i - \int\limits_{N+1}^{\infty} f(x)\,dx \right| \ \leq\ \int\limits_{N}^{N+1} f(x)\,dx$$

Noting that $\sum_{i=N+1}^{\infty} a_i = \sum_{i=1}^{\infty} a_i - \sum_{i=1}^{N} a_i$, we have the inequality that forms the basis for our approximation method

$$\left| \sum_{i=1}^{\infty} a_i - \left(\sum_{i=1}^{N} a_i + \int_{N+1}^{\infty} f(x)\,dx \right) \right| \leq \int_{N}^{N+1} f(x)\,dx \leq a_N$$

The last part of the inequality results from the fact that f is a decreasing function, the width of the interval $[N, N+1]$ is 1, and $f(N) = a_N$. This inequality translates to the following approximation method.

If $f(x)$ is a nonnegative decreasing function with $f(n) = a_n$ such that $\int_{1}^{\infty} f(x)\,dx$ converges, then the series $\sum_{i=1}^{\infty} a_i$ may be approximated by

$$\sum_{i=1}^{N} a_i + \int_{N}^{\infty} f(x)\,dx .$$

The error of this approximation to the series does not exceed a_N .

As an illustration of this method, let's approximate the value of the series $\sum_{i=1}^{\infty} \frac{1}{i^2}$. We noted in the introduction that the sum of the first 500 terms is 1.64293 . We use DERIVE to approximate the value of the integral $\int_{501}^{\infty} \frac{1}{x^2}\,dx$. This is done by first creating the expression

$$\int_{501}^{t} \frac{1}{x^2}\,dx .$$

and simplifying. Apply the Calculus/Limit option to take the limit of the result as $t \to \infty$. Approximating this result yields .001996 . Adding this to the previously obtained sum of the first 500 terms gives the approximation of 1.644926 for the value of the series. This approximation is in error by at most $\frac{1}{500^2} = .000004$. Thus, we know that the value of the series lies between 1.644922 and 1.644930 .

What accuracy can be obtained by summing 1000 terms of this series and using the approximation method developed above? Use DERIVE to find this approximation. Was the increased accuracy worth the extra computation? How many terms of this series need to be summed to guarantee that the first nine decimal places of the approximation are correct?

An Adaptation of the Ratio Test

There are certain series that DERIVE can evaluate directly. One class of such series is that of the geometric series, i.e., series of the form $\sum_{i=1}^{\infty} r^i$. For example , author $\left(\frac{1}{3} \right)^i$ and use the

Calculus/Sum option with an upper limit of "inf" to create the expression, $\sum_{i=1}^{\infty} \left(\frac{1}{3}\right)^i$.   Simplifying

this expression yields a result of $\frac{1}{2}$, the sum of the series.   As a further test of the ability of DERIVE

to evaluate this type of series, create the expression $\sum_{i=1}^{\infty} r^i$ and evaluate it for several choices of $r > 0$.

What happens when $r \geq 1$?   What is the sum of the series for any positive $r < 1$?

Now, let's turn to a series whose terms, a_i, are positive, decreasing, and satisfy the following
condition

$$\lim_{i \to \infty} \frac{a_{i+1}}{a_i} = \gamma < 1 .$$

Such a series satisfies the hypotheses of the ratio test and, hence, we know that the series converges. If

the a_i satisfy the additional property that the sequence $\left\{\frac{a_{i+1}}{a_i}\right\}$ is decreasing, then the fact that the

limit of the sequence is less than 1 has the consequence that for some integer N,

$$\frac{a_{N+1}}{a_N} = r < 1.$$

But then $a_{N+1} = r \cdot a_N$.   Since, the sequence of ratios is decreasing, we have

$$\frac{a_{N+2}}{a_{N+1}} < \frac{a_{N+1}}{a_N} = r$$

Thus,

$$a_{N+2} < r \cdot a_{N+1} = r^2 a_N$$

Likewise,

$$a_{N+3} < r \cdot a_{N+2} = r^3 a_N$$

$$\cdot \qquad \cdot \qquad \cdot$$
$$\cdot \qquad \cdot \qquad \cdot$$
$$\cdot \qquad \cdot \qquad \cdot$$

$$a_{N+k} < r \cdot a_{N+k-1} = r^k a_N$$

Summing all of the terms of the sequence for $i \geq N$, we have

$$\sum_{i=N+1}^{\infty} a_i < \sum_{i=1}^{\infty} r^i a_N = \frac{r \cdot a_N}{1-r} .$$

This gives us a bound on the tail end of the series and, hence, a bound on the error from summing N

terms of the sequence to estimate the value of the series.

If the sequence $\{a_n\}$ satisfies the conditions given in this section, then the error

from estimating the value of the series $\sum_{i=1}^{\infty} a_i$ by the sum $\sum_{i=1}^{N} a_i$ does not exceed

$\frac{r \cdot a_N}{1-r}$, where $r = \frac{a_{N+1}}{a_N}$.

As an example of this observation, suppose that we wish to estimate the value of the series,

$$\sum_{i=1}^{\infty} \frac{1}{i!}$$

to a value within 10^{-6} of the actual value of the series. The result we have developed tells us that we must find an N such that

$$\frac{r \cdot a_N}{1-r} \leq 10^{-6}$$

where

$$r = \frac{\frac{1}{(N+1)!}}{\frac{1}{N!}} = \frac{1}{N+1}$$

Substituting for r and a_N in the inequality,

$$\frac{\frac{1}{N+1} \cdot \frac{1}{N!}}{1 - \frac{1}{N+1}} = \frac{1}{N \cdot N!} < 10^{-6}$$

or

$$1{,}000{,}000 < N \cdot N! \ .$$

Using DERIVE's factorial function, we see that when $N = 9$, $N \cdot N! = 3{,}265{,}920$ and summing nine terms of the sequence will give us the desired estimate. This is a rather high degree of accuracy from summing only a few terms of the sequence. The value of the sum of the first nine terms is 1.718281 .

The Laboratory Report

Choose the appropriate Precision setting and determine the value of N to determine the value of the series $\sum_{i=1}^{\infty} \frac{1}{i!}$ with an error of no more than 10^{-10}. Show all of your work to justify your choice of N. What is the value of the estimate?

Determine that each of the following series converges and then use the appropriate method from this lab to estimate the value of the series with an error of no more than 10^{-6}.

$$\sum_{i=2}^{\infty} \frac{1}{i \, (\ln i)^2} \qquad\qquad \sum_{i=1}^{\infty} \frac{\operatorname{atan} i}{1 + i^2}$$

$$\sum_{i=1}^{\infty} \frac{2^i}{i!} \qquad\qquad \sum_{i=1}^{\infty} \frac{1}{\sqrt{n^3}}$$

$$\sum_{i=1}^{\infty} \frac{1 \cdot 3 \cdot 5 \, \cdots \, (4i - 3)}{n!}$$

Work Sheet

Laboratory #16

1. How many terms of the series $\sum_{i=1}^{\infty} \frac{1}{i!}$ are necesary to give an estimate with an error of no more than 10^{-10}?

2. Fill in the table.

Series	Converge?	# Terms (accuracy: 10^{-6})	Value
$\sum_{i=2}^{\infty} \dfrac{1}{i(\ln i)^2}$			
$\sum_{i=1}^{\infty} \dfrac{2^i}{i!}$			
$\sum_{i=1}^{\infty} \dfrac{\arctan i}{1 + i^2}$			
$\sum_{i=1}^{\infty} \dfrac{1}{\sqrt{n^3}}$			
$\sum_{i=1}^{\infty} \dfrac{1 \cdot 3 \cdot 5 \cdots (4i - 3)}{n!}$			

Laboratory #17

Using Polynomials to Approximate the Values of Functions

Introduction

How does one develop routines to quickly evaluate functions such as the sine, cosine, natural logarithm, or exponential function? It is impractical to store a huge table of values in a calculator or even in our own memory. The solution is to store a few selected values of the function and routines that efficiently approximate the other values. Most of the approximation routines include evaluating a polynomial that closely approximates the function. For example, consider the graph of x and sin(x) over the interval $(-\pi, \pi)$ as shown in Figure 17-1. These two functions overlap in the interval $(-.2, .2)$; however, when they part company — never the twain shall meet again. Nonetheless, in the region from $-.2$ to $.2$, the value of x may be a good approximation to x. As another approximation, consider the graphs of sin(x) and $x - \dfrac{x^3}{6}$. For this cubic polynomial, the interval of overlap with sin(x) is even wider! It appears to be from $-.7$ to $.7$. Once again, for x outside of this interval, the graphs do not agree at all.

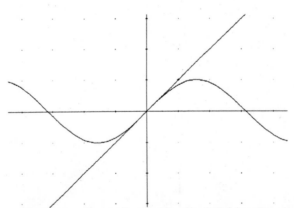

Figure 17-1: Comparing the Graphs of sin(x) and x

The question we will consider in this lab is How does one come up with the polynomials that approximate the functions and for what region is the approximation valid? Before we tackle the general question, we will investigate the behavior of polynomials themselves.

Determining a Polynomial from the Values of Its Derivatives at a Point

Let's start with a general quadratic polynomial, $p(x) = \alpha_0 + \alpha_1(x-a) + \alpha_2(x-a)^2$ and assume that we know the values of $p(a)$, $p'(a)$, $p''(a)$. Can we find α_0, α_1, α_2 ? Note that there is no loss of generality in writing the polynomial in powers of $(x-a)$ instead of powers of x. Doing this makes our computations easier and, quite possibly, the discussion more clear.

Finding α_0 is easy: $p(a) = \alpha_0 + \alpha_1(a-a) + \alpha_2(a-a)^2 = \alpha_0$. Thus, we know that $p(x) = p(a) + \alpha_1(x-a) + \alpha_2(x\text{-}a)^2$. It remains to determine α_1 and α_2. This is almost as easy as finding the value for α_0. Taking the derivative of $p(x)$ drops the constant term and yields a linear polynomial in $(x-a)$ with α_1 as the constant term.

$$p'(x) = \alpha_1 + 2\cdot\alpha_2(x-a)$$

Proceeding as before, we see that $p'(a) = \alpha_1$ and $p(x) = p(a) + p'(a)\cdot(x-a) + \alpha_3(x-a)^2$. Taking the second derivative of p, we have

$$p''(x) = 2\cdot\alpha_2$$

and, thus, $\alpha_2 = \dfrac{p''(a)}{2}$. We have completely determined $p(x)$:

$$p(x) = p(a) + p'(a)\cdot(x-a) + \frac{p''(a)}{2}\cdot(x-a)^2 .$$

As an application of the above result, suppose that a ball is thrown upward from a height of 7 feet with a velocity of 60 ft/sec. What is the position of the ball t seconds after it is released?

If $s(t)$ is the position of the ball at time t, then we know $s(0) = 7$, $s'(0) = 60$, and (courtesy of Newton) $s''(0) = -32$. Thus, we immediately have
$$s(t) = 7 + 60t - 16t^2 .$$

In order to extend the above argument to a general polynomial of degree N,

$$p(x) = \alpha_0 + \alpha_1(x-a) + \alpha_2(x-a)^2 + \cdots + \alpha_N(x-a)^N$$

we observe that if $p^{(k)}(x)$ denotes the kth derivative of $p(x)$ with respect to x, then

$$p^{(k)}(x) = k\cdot(k\text{-}1)\cdots 3\cdot 2\cdot 1\,\alpha_k + \text{remaining terms involving powers of } (x-a).$$

Thus,

$$p^{(k)}(a) = k\cdot(k\text{-}1)\cdots 3\cdot 2\cdot 1\,\alpha_k = k!\,\alpha_k$$

and we have

$$\alpha_k = \frac{p^{(k)}(a)}{k!}$$

This means that

$$p(x) = p(a) + p'(a)\cdot(x-a) + \frac{p''(a)}{2}\cdot(x-a)^2 + \cdots + \frac{p^{(N)}(a)}{n!}\cdot(x-a)^N$$

and we can completely determine a polynomial of degree N by knowing its value and the value of its N derivatives at some x = a. The coefficient of $(x-a)^k$ is $\dfrac{p^{(k)}(a)}{k!}$.

Polynomial Approximations of sin(x) and cos(x)

In this section we will use the fact that if $f(x) \leq g(x)$ on some interval [a, b] and both functions are integrable on this interval, then

$$\int_a^b f(x)\,dx \; \leq \; \int_a^b g(x)\,dx \; .$$

How this fact leads to a polynomial approximation of the sine and cosine is rather interesting. The inequality lets us trap these functions between two polynomials. We begin by noting that $\cos(x) \leq 1$ for all x. Thus,

$$\int_0^x \cos(x)\,dx \; \leq \; \int_0^x dx$$

or,

$$\sin(x) \; \leq \; x \qquad \text{for all } x \geq 0$$

We integrate both sides of this inequality from 0 to x and see that

$$-\cos(x) + 1 \; \leq \; \frac{x^2}{2} \qquad \text{for all } x \geq 0$$

But since $\cos(x) \leq 1$ for all x, we can extend this inequality to

$$0 \; \leq \; -\cos(x) + 1 \; \leq \; \frac{x^2}{2}$$

Multiplying by -1 and adding 1 to all parts of the inequality,

$$1 - \frac{x^2}{2} \; \leq \; \cos(x) \; \leq \; 1$$

we integrate all parts of this inequality from 0 to x to obtain an inequality for sin(x)

$$x - \frac{x^3}{6} \; \leq \; \sin(x) \; \leq \; x$$

Continuing in this fashion, we develop the following string of inequalities:

$$1 - \frac{x^2}{2} \; \leq \; \cos(x) \; \leq 1 - \frac{x^2}{2} + \frac{x^4}{24}$$

$$x - \frac{x^3}{6} \; \leq \; \sin(x) \; \leq x - \frac{x^3}{6} + \frac{x^5}{120}$$

$$1 - \frac{x^2}{2} + \frac{x^4}{24} - \frac{x^6}{720} \; \leq \cos(x) \; \leq 1 - \frac{x^2}{2} + \frac{x^4}{24}$$

$$x - \frac{x^3}{6} + \frac{x^5}{120} - \frac{x^7}{5040} \leq \sin(x) \leq x - \frac{x^3}{6} + \frac{x^5}{120}$$

You should supply the steps necessary to generate these inequalities, i.e., the integration and algebra. We make the following observations from studying the above sequence of inequalities.

1. For each term of the listed polynomials, the denominator is the factorial of the exponent.

2. The signs of the terms of the polynomial alternate.

3. The polynomials on either side of the trigonometric function differ by one term of the form $\frac{x^k}{k!}$.

We deal with observation 3 first. The term $\frac{x^k}{k!}$ becomes small as k becomes large for any value of x. To illustrate this point, Figure $17-2$ is the graph of $\frac{x^{10}}{10!}$. Note that in the range of $-\pi$ to π the graph is barely distinguishable from the x-axis. This means that a polynomial of relatively small degree can be used to approximate the sine and cosine functions on this interval.

Figure 17-2: An Illustration of the Dominance of the Factorial over the Exponential

We illustrate the convergence of the sequence of polynomials generated to approximate the sine function by graphing sin(x) and the seventh-degree polynomial that we generated as part of our analysis. The graphs of these two functions are shown in Figure $17-3$.

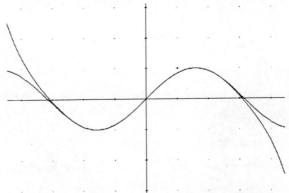

Figure 17-3: Sin(x) and a Seventh-Degree Polynomial Approximation

Observations 1 and 2 are even more interesting in light of the previous section on polynomials. Note that if we evaluate the derivatives of cos(x) at x = 0, we get the sequence 1, 0, − 1, 0, 1, 0, − 1, 0, etc. , and the derivatives of sin(x) at x = 0 yield the sequence 0, 1, 0, − 1, 0, 1, 0, − 1, 0, etc. This means that in the polynomials for sin(x) and cos(x) have the property that the coefficient of the kth term is given by

$$\frac{f^{(k)}(0)}{k!} \qquad \text{where f(x) = either sin(x) or cos(x) .}$$

This result is amazing when we think of how the polynomial approximations were generated. In fact, it is too amazing to attribute to coincidence. We will see that this occurrence goes far beyond the sine and cosine functions.

Taylor Polynomials

Suppose that the first N derivatives of f(x) exist at x = a; then the Taylor Polynomial of degree N for f(x) at x=a is defined as

$$P(x) = f(a) + f'(a) \cdot (x - a) + \frac{f''(a)}{2} \cdot (x - a)^2 + \cdots + \frac{f^{(N)}(a)}{N!}$$

In the previous section we developed Taylor Polynomials of degree 6 and 7 about 0 for cosine and sine, respectively. The above expression gives a more direct way of developing these polynomials by taking the derivatives and evaluating them at x = a, which is 0 in the case of the approximations for sine and cosine.

DERIVE will generate Taylor Polynomials as part of its Calculus menu. To demonstrate, enter sin(x) and press "c" for Calculus. Pressing "t" for Taylor starts a series of questions. Press return to indicate that you are seeking a Taylor Polynomial for the highlighted expression. Next enter 0 for the point and 7 for the degree. The expression TAYLOR(sin(x),0,7) will appear on the screen.

Simplifying this expression will produce the polynomial we developed above.

We can, of course, apply this operation to other functions. We can estimate the interval for which the polynomial approximates the function by graphing. If we draw the graph of e^x and its 10th degree Taylor Polynomial over the region $[-4,4]$, we see that in this interval the fit is almost perfect.

This is not always the way things work. Figure $17-4$ shows the graph of $f(x) = \ln(x)$ and its 10th-degree Taylor Polynomial about $x = 1$. The fit on the interval of agreement is quite good, but you should use DERIVE to verify that increasing the degree of the polynomial will not expand the interval in any significant way. Why do you think this is the case? Look at the terms of the series. What happens when $x \geq 2$? In your laboratory report you will be investigating the relationship between Taylor Polynomials and the functions they approximate. The interval of agreement is important for determining for what values of x an approximation will be valid.

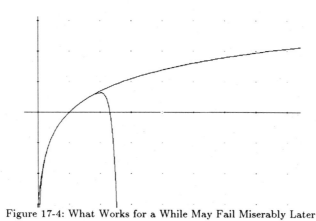

Figure 17-4: What Works for a While May Fail Miserably Later

The Laboratory Report

Write a description of the process of approximating a function by using a Taylor Polynomial. Discuss the motivation for the idea of a Taylor Polynomial and develop formulas for a general polynomial of degree n for $\cos(x)$, $\sin(x)$, and e^x about $x = 0$ and $\ln(x)$ about $x = 1$. Discuss your observations of the interval of agreement for each of the polynomials and the functions used to generate the polynomials.

Extra Credit

Consider the graphs of the functions $\sin(x)$ and $x*(\cos(x))^{\frac{1}{3}}$. Note that they agree rather well on the interval $(-1, 1)$ and very poorly outside of this interval. Use DERIVE to generate 11th-degree Taylor Polynomials for these functions about $x = 0$. What do you notice about these polynomials? Does this explain why they seem to agree on $(-1, 1)$ and disagree wildly elsewhere? [6]

---

[6]This example was suggested by Jerry Uhl of the University of Illinois.

Work Sheet

Laboratory #17

1. For each of the following use DERIVE to find the fifth degree Taylor Polynomial about the indicated point.

   a. $\sin(x)$ about $x = 0$

   b. $\cos(x)$ about $x = 0$

   c. e^x about $x = 0$

   d. $\ln(x)$ about $x = 1$

2. For each of the above, draw the graph of the function and the 5th-degree Taylor Polynomial. Use the graph to make an estimate the largest interval on which you think the polynomial approximation will be accurate.

   a. $\sin(x)$

   b. $\cos(x)$

   c. e^x

   d. $\ln(x)$

Laboratory #18

Applying Polynomial Methods

Introduction

In Lab 17 we saw the motivation for generating the Taylor Polynomials for a function, $f(x)$, and examined graphical evidence for the agreement between the values of a Taylor Polynomial for $f(x)$ at $x = a$ and the values for $f(x)$ in a neighborhood of $x = a$. This agreement makes it possible to estimate particular values of $f(x)$ with a good degree of accuracy. In this lab we will examine the use of Taylor polynomials to estimate the numerical values of integrals and the solutions to differential equations.

For example, suppose that we have a need to evaluate the integral

$$\int_0^{\sqrt{\pi/2}} \sin(x^2)\,dx \ .$$

We know of no function whose derivative is $\sin(x^2)$. Consequently, we cannot evaluate this integral by using the Fundamental Theorem of Calculus. We need to use some approximation technique.

Expanding the 10th-degree Taylor Polynomial for $\sin(x^2)$ about $x = 0$, we have:

$$x^2 - \frac{x^6}{6} + \frac{x^{10}}{120} \ .$$

The strategy will be to integrate the series and substitute the value of this integral as an approximation to the value of the original integral. How accurate is this approximation? In Figure $18-1$ we see the graphs of $\sin(x^2)$ and the 10th-degree Taylor Polynomial for $\sin(x^2)$ for x in the interval from 0 to 2.

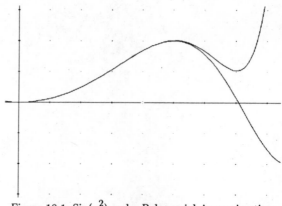

Figure 18-1: $\text{Sin}(x^2)$ and a Polynomial Approximation

page 130

The two graphs almost coincide in the interval from 0 to $\sqrt{\pi/2}$. Outside of this interval they are quite different. However, since the graphs virtually coincide from 0 to $\sqrt{\pi/2}$, the area under the graphs must be very close to equal. It is very easy to evaluate

$$\int_0^{\sqrt{\pi/2}} \left(x^2 - \frac{x^6}{6} + \frac{x^{10}}{120} \right) \, dx$$

quite accurately. Using DERIVE, we evaluate the integral as .549657. Thus, we conclude that

$$\int_0^{\sqrt{\pi/2}} \sin(x^2) \, dx \approx .549657$$

In this instance we used the graphs of the function and its Taylor Polynomial to justify the accuracy of the approximation. As we will see in the next example, it is not always possible to graph the function, and more reliable methods will need to be developed to justify the accuracy of the estimate. The development of these methods is a topic for a course in numerical analysis. We will take on faith the accuracy of our approximations.

Using Taylor Polynomials to Estimate the Solution of a Differential Equation

In their 1973 book[7], Baxter and Sloyer propose the following model for the national debt, based on national income

Suppose that national income, I(t), and national debt, D(t), both increase at a rate proportional to the national income. That is,

$$\frac{dI}{dt} = rI \qquad \text{and} \qquad \frac{dD}{dt} = sI$$

where r and s are constants.

If we substitute $I = \frac{1}{s}\frac{dD}{dt}$ in the first equation in the above statement, we have

$$\frac{1}{s}\frac{d^2D}{dt^2} = \frac{r}{s}\frac{dD}{dt}$$

or

$$\frac{d^2D}{dt^2} - r\frac{dD}{dt} = 0 \ .$$

Let's suppose that $r = s = .02$ and that at the present time, $t = 0$, $D = \$4.5$ billion and $I = \$7$ billion.

---

[7]Baxter,W.E. & Sloyer, E.W., *Calculus with Probability for the Life and Management Sciences,* Addison-Wesley, Reading, Mass.: 1973, page 465.

As a technique for investigating the nature of the solution to the second order differential equation given above, assume that we can expand D as a series about $t = 0$, i.e.,

$$D(t) \ = \ \sum_{i=0}^{\infty} a_i t^i \ .$$

If we furthermore assume that we are working within the radius of convergence for this series, we can substitute the series for D in the differential equation. Before doing this, note the following (pay particular attention to the index, subscripts, and superscripts)

$$\frac{dD(t)}{dt} \ = \ \sum_{i=1}^{\infty} i \cdot a_i t^{i-1} \ = \ \sum_{i=0}^{\infty} (i+1) a_{i+1} t^i$$

$$\frac{d^2D(t)}{d^2t} \ = \ \sum_{i=1}^{\infty} i \cdot (i+1) a_{i+1} t^{i-1} \ = \ \sum_{i=0}^{\infty} (i+1)(i+2) a_{i+2} t^i$$

Substituting these series into the differential equation (recall $r = .02$)

$$\sum_{i=0}^{\infty} (i+1)(i+2) a_{i+2} t^i \ - .02 \sum_{i=0}^{\infty} (i+1) a_{i+1} t^i \ = 0$$

and combining terms yields the equation

$$\sum_{i=0}^{\infty} (i+1) \big[(i+2)a_{i+2} - .02 \cdot a_{i+1} \big] t^i \ = \ 0$$

This sequence of equations may seem intimidating at first, but it is really nothing more than an extension of the ideas of using a polynomial to approximate $D(t)$. If you have trouble following the above argument, assume

$$D(t) \ \approx \ a_0 \ + \ a_1 t \ + \ a_2 t^2 \ + \ a_3 t^3 \ + \ a_4 t^4 \ + \ a_5 t^5$$

and substitute this into the differential equation. Look at the coefficients of t, t^2, t^3, and the constant coefficient.

In any case, since the series equals 0 for all t, the coefficients of the series in the above equation must all be 0. Thus,

$$(i+2)a_{i+2} \ - \ .02 \cdot a_{i+1} \ = \ 0 \quad \text{for all } i = 0, 1, 2, \ldots$$

Substituting $i - 1$ for i, we have

$$(i+1)a_{i+1} \ - \ .02 \cdot a_i \ = \ 0 \quad \text{for all } i = 1, 2, 3, \ldots$$

This gives the recursive relationship

$$a_{i+1} \ = \ \frac{.02}{i+1} a_i \quad \text{for all } i = 1, 2, 3, \ldots$$

Starting with a_2 and generating a few terms of this sequence, we see that

$$a_{i+1} = \frac{(.02)^i}{(i+1)!} a_1 \qquad \text{for all } i = 1, 2, 3, \ldots$$

or, replacing $i + 1$ by i,

$$a_i = \frac{(.02)^{i-1}}{i!} a_1 \qquad \text{for all } i = 2, 3, 4, \ldots$$

We know from our initial information that $D(0) = 4.5$ and $\frac{dD(0)}{dt} = .02 * 7 = .14$. Substituting $t = 0$ in the series for D we have

$$D(0) = a_0 + \sum_{i=1}^{\infty} a_i \, 0^i = a_0 = 4.5$$

$$\frac{dD(0)}{dt} = a_1 + \sum_{i=2}^{\infty} (i+1) \, a_i \, 0^i = a_1 = 7 * .02$$

Combining all of these observations, we finally have

$$D(t) = 4.5 + \sum_{i=1}^{\infty} \frac{(.02)^{i-1}}{i!} .02 * 7 \; t^i$$

or

$$D(t) = 4.5 + 7 \sum_{i=1}^{\infty} \frac{(.02)^i}{i!} t^i$$

This technique has its moments of tedious bookkeeping, but it pays off in giving us a series solution to the differential equation. The graph of $D(t)$ is given in Figure $18 - 2$.

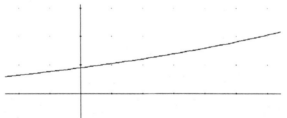

Figure 18-2: The Solution of the Differential Equation Model for the National Debt

The fundamental assumption that we made was that the solution to the differential equation had an expansion as an infinite series. The most difficult part of implementing this technique, if it is applicable, is manipulating the indices of the terms of the series and the range of the index. A standard operating procedure is to make the adjustments so that the exponent of the independent variable is merely the index. At times this may mean allowing a few of the leading terms of the series to stand alone outside of the series. This was the case with the a_0 and a_1 terms in the above example.

The Laboratory Report

The above technique can be applied to differential equations having nonconstant coefficients. Apply this method to solve the differential equation

$$y'' - x\,y' - y = 0$$

satisfying the initial conditions $y(0) = 0$, $y'(0) = 1$.

In your report explain how you arrived at your solution and include a graph of the 10th-degree polynomial that approximates this solution.

Work Sheet

Laboratory #18

1. Determine the degree of a Taylor Polynomial to approximate the value of the integral

$$\int_0^1 e^{-x^2}\, dx$$

    What is your approximation?

2. Use the methods described in the manual to find a series about $x = 0$ that satisfies the differential equation

$$y'' - x\,y' - y = 0$$

    with $y(0) = 0$ and $y'(0) = 1$.

Laboratory #19

Exploring Polar Coordinates

Introduction

When a pilot or a ship's navigator requests directions to a destination, the information is given in terms of a heading (angle) and a distance. Although grid, or Cartesian, coordinates can be extremely useful for locating a position on a map, they are not immediately useful in determining a course for vehicles that have complete two-dimensional freedom of motion. There is even evidence that when bees are communicating the location of a flower garden, they communicate a direction and a distance to the rest of the swarm by doing a dance that gives these two pieces of information.[8] This method of describing the location of a point in the plane is a natural and easy method of communication.

In order to understand this method of labeling a point in the plane, consider a point with Cartesian coordinates, (x, y).

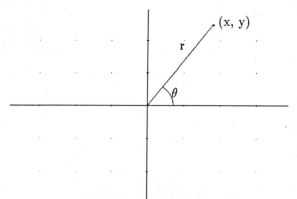

Figure 19-1: Describing a Point in Polar Coordinates

From this point draw a line to the origin. This line makes an angle, θ, with the x-axis and is a distance, r, from the origin. We will use these two values (r, θ) to describe the point. Thus, we have

$$r = \sqrt{x^2 + y^2}$$

and the angle θ is determined by the fact that

---

[8]See, for example, E. Batschelet, *Introduction to Mathematics for Life Scientists*, Springer-Verlag, New York, N.Y.: 1973, §5.3

$$x = r\cos(\theta) \quad \text{and} \quad y = r\sin(\theta).$$

There are some problems with this definition. For example, if we do not restrict θ, the description of the point is not unique, since

$$r\cos(\theta) = r\cos(\theta + 2n\pi) \quad \text{and} \quad r\sin(\theta) = r\sin(\theta + 2n\pi)$$

for any integer n. This problem is solved by restricting the angle, θ, used in the description of a point to lie in the interval $-\pi < \theta \le \pi$ radians. If the point is described using an angle outside of these bounds, it is reduced modulo 2π (i.e., add or subtract 2π until the angle lies between $-\pi$ and π). This establishes a unique description for all points save one. The origin can be described by $r = 0$ and any angle, θ. Generally, this problem is ignored by simply describing the origin as $r = 0$ and not specifying an angle or also specifying it as 0.

DERIVE has the capability to give the polar description of points in the plane. Operate in the graphics window by pressing "p" for Plot. Now press "o" for Options and "t" for Type. The word Rectangular should be highlighted. Press "p" to highlight the word Polar. Press return and note the description of the cross hair position in the bottom left hand corner of the screen. Instead of describing the position in (x, y) coordinates, it changes to r and θ. To get a feel for polar coordinates, move the cross hair to several familiar positions on the plane and note the polar coordinates. What is the polar description for the origin chosen by DERIVE? Another interesting experiment is to move the cross hair to the third quadrant and slowly step it across the negative x-axis. Note that the value for θ takes a jump of 2π from π to $-\pi$! This type of behavior becomes very important in the study of functions of a complex variable. At this point we will only note that the polar coordinate description of the plane yields this unusual behavior along the negative x-axis.

Plotting Polar Equations

It is always possible to convert a Cartesian (x,y) equation into a polar equation by substituting the expression $r\cos(\theta)$ for x and $r\sin(\theta)$ for y. If it is possible to solve for r, then we have r as a function of θ. In this case our test for a function is that for each θ there is only one corresponding value of r. Let's start with a simple example. We know that $x^2 + y^2 = 4$ is a circle of radius 2 centered at the origin. Making the substitution for x and y yields,

$$r^2 = 4$$

or

$$r = 2 \quad \text{and} \quad r = -2.$$

Since we interpret r as a distance, we accept the first solution and reject the second answer. Authoring the expression, 2, and plotting in polar coordinates, choosing to let θ vary between $-\pi$ and π, we see the graph of a circle traced out in a counter-clockwise direction. This is to be expected. Note that if you choose to extend the domain of θ beyond $-\pi$ to π, the graph merely wraps over itself. This is, of course, due to the periodic behavior of the sine and cosine functions.

Now let's experiment some more. Return to the algebra window and author the expression, -2. Clear the graphics window and attempt to plot this expression. You should be surprised to note that it also plots a circle, but it does it differently. It plots in a counter-clockwise direction. The explanation for this is due to the trigonometric identities

$$-\sin(\theta) = \sin(\theta + \pi) \quad \text{and} \quad -\cos(\theta) = \cos(\theta + \pi) .$$

If we convert $-2\sin(\theta)$ to $2\sin(\theta+\pi)$ and $-2\cos(\theta)$ to $2\cos(\theta+\pi)$ we find that the rather strange polar coordinates of $(-2, \theta)$ are transformed to the acceptable form $(2, \theta+\pi)$. Thus, we can graph the polar function r:$= -2$. We use this convention of augmenting the angle by π whenever we are faced with the situation of plotting a point with r < 0.

Let's look at some of the standard polar equations. The first is the famous *cardioid*, so named for its shape, which is reminiscent of a heart. The polar equation of this curve is given by

$$r:= a(1 + \cos(\theta)) .$$

We choose a $= 2$ and author the expression $2(1 + \cos(\theta))$. Note: θ is authored as Alt-h in DERIVE.

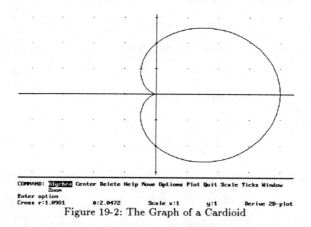

COMMAND: Algebra Center Delete Help Move Options Plot Quit Scale Ticks Window
Zoom
Enter option
Cross r:1.0901   θ:2.0472   Scale x:1   y:1   Derive 2D-plot

Figure 19-2: The Graph of a Cardioid

The plot in Figure $19-2$ is a graph of this curve. Note its shape. What is the value for r when

$\theta = -\pi, \ -\frac{\pi}{2}, \ 0, \ \frac{\pi}{2}, \ \pi$? Now return to the Algebra window and author the expression

$2(1 + \cos(\theta+\frac{k\pi}{2}))$. Use the Manage/Substitute option to substitute k $= 1, \ 2, \ 3$ and plot the

expression for each of these choices. How does the choice of k affect the graph?

The graphs of the curves r:$= a \cdot \sin(\theta)$ and r:$= a \cdot \cos(\theta)$ are rather surprising. What curves do they generate? How many times are these curves traced? Plot the first one for θ ranging from 0 to π and the second one for θ ranging from $\frac{-\pi}{2}$ to $\frac{\pi}{2}$. What role is played by the parameter a?

The graphs of the curves r:= $a \cdot \sin(k\theta)$ for k = 2, 3, ... are called *rosettes*. Set a = 1 and draw the graphs for k = 2, 3, 4, 5. Notice that there is a different behavior when k is even and when k is odd. Describe this behavior for the cases you drew and make a general conjecture about the graphs.

Laboratory Report

Consider the class of polar equations of the form

$$r:= \frac{1}{1 + e\cos(\theta)}$$

for several values of e. In particular, consider e in the following ranges

$$e = 0$$
$$0 < e < 1$$
$$e = 1$$
$$e > 1 \,.$$

Are there values of θ that must be avoided in any of the above? What is the role of e?

As a further investigation, determine the behavior of the curves if you change $\cos(\theta)$ to $\sin(\theta)$. Also, what happens if you change the + to −? Recall that $\sin(\theta) = \cos(\theta + \frac{\pi}{2})$, $-\cos(\theta) = \cos(\theta + \pi)$, and $-\sin(\theta) = \cos(\theta + \frac{3\pi}{2})$. In general, give a complete description of the graph of the polar equation

$$r:= \frac{1}{1 + e\cos(\theta + \alpha)}$$

where α is fixed. Pay particular attention to the role of e and α.

Work Sheet

Laboratory #19

1. Draw the polar graphs of

$$2\left(1 + \cos\left(\theta + \frac{k\pi}{2}\right)\right)$$

for $k = 1, 2, 3$.

2. a. Draw the polar graph of $2\sin\theta$ for $0 \leq \theta \leq \pi$.

 b. Draw the polar graph of $2\cos\theta$ for $\frac{-\pi}{2} \leq \theta \leq \frac{\pi}{2}$.

3. Draw the polar graphs of $\sin(k\theta)$ for $0 \leq \theta \leq 2\pi$ and $k = 2, 3, 4, 5$.

Laboratory #20
Functions of More Than One Variable

Introduction

When we consider many applications of mathematics, we need to go beyond the situation where two variables, one dependent and one independent, are sufficient to model the situation. For example, to determine the optimum production schedule for a firm manufacturing three products it is necessary to consider the profit resulting from the manufacture of the product as a function of three variables. In this lab we will consider techniques to solve such problems as an extension of the techniques used to solve problems for functions of one variable. This will only serve as an introduction to the subject and leave the study of more inclusive techniques to a course in multivariable calculus.

We begin by considering a function of two variables, x and y. Suppose

$$f(x, y) = 3 - \tfrac{1}{4}(x^2 + y^2 - 2x - 4y).$$

A plot of the surface defined by setting $z = f(x, y)$ is given in Figure $20 - 1$.

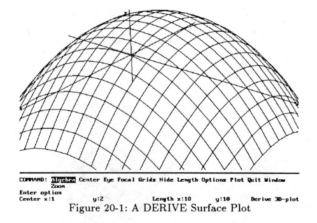

COMMAND: **Algebra** Center Eye Focal Grids Hide Length Options Plot Quit Window
Zoom
Enter option
Center x:1   y:2   Length x:10   y:10   Derive 3D-plot

Figure 20-1: A DERIVE Surface Plot

It is clear from this figure that there is a point (x_0, y_0) for which $f(x, y)$ attains a maximum value, i.e., there is a highest point on the surface. This maximum value, since we know that it exists, can be located by defining two functions of one variable and locating the maxima of these functions.

Define $F(x) = f(x, y_*)$ and $G(y) = f(x_*, y)$ where x_* and y_* are assumed to be constants. In other words, $F(x)$ is a function of x alone obtained from f by keeping y constant, and $G(y)$ is a function of y alone obtained from f by keeping x constant. We then approach the problem as one of maximizing two functions of one variable, i.e., we search for (x_0, y_0) such that $F(x_0)$ is the maximum value for F and $G(y_0)$ is the maximum value for G. Thus, we consider all x_0 and y_0 that are critical points for F and G, respectively, or locate x_0 and y_0 such that

$$\frac{df(x, y_0)}{dx} = 0 \quad \text{and} \quad \frac{df(x_0, y)}{dy} = 0 \; .$$

In multivariable calculus courses the notation for the derivatives is simplified by using a partial derivative symbol, and the above equations are written as

$$\frac{\partial f(x, y)}{\partial x} = 0 \quad \text{and} \quad \frac{\partial f(x, y)}{\partial y} = 0.$$

The DERIVE keystrokes correspond to this notation.

Let's use DERIVE to find the maximum value for $f(x, y)$, which we know exists from the graph. After entering the expression for $f(x, y)$, use the Calculus/Differentiate operator to take the derivative with respect to the x-variable. Highlight the expression for $f(x, y)$ again and use the Calculus/Differentiate operator. This time choose y as the differentiate variable. Form a system of equations with these two derivatives and solve. The result is $x = 1$ and $y = 2$.

We can use the same technique to find the location of a local minimum, if we know it exists. In effect all we are doing is using our experience from functions of one variable and extending it to the multivariable situation. Some words of warning: Do not blindly rely on the above tests without first obtaining evidence that either a maximum or minimum exists. Below is the plot of the surface obtained from a nice, seemingly benign, function $f(x, y) = x^2 - y^2$.

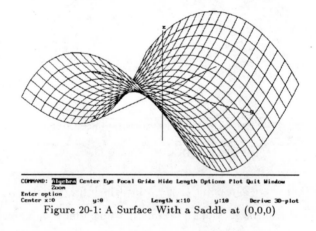

COMMAND: **Algebra** Center Eye Focal Grids Hide Length Options Plot Quit Window
Zoom
Enter option
Center x:0   y:0   Length x:10   y:10   Derive 3D-plot

Figure 20-1: A Surface With a Saddle at (0,0,0)

For this surface, note that when $x = 0$ and $y = 0$, $f(x, 0)$ has a minimum and $f(0, y)$ has a maximum. This creates the saddle effect. Thus, the condition

$$\frac{\partial f(x, y)}{\partial x} = 0 \quad \text{and} \quad \frac{\partial f(x, y)}{\partial y} = 0$$

is a necessary, but not sufficient, condition for a differentiable function to have a local maximum or minimum at a particular point.

Test the following functions for the existence of maxima or minima by doing a surface plot using DERIVE. If there is evidence of the existence of such a point, locate the maximum or minimum point on the surface.

1. $2x - x^2 - y^2$
2. $x^2 + xy + 2y^2 - 3x + 2y$
3. $2x^2 + xy - y^2$
4. $(x^2 + y^2) e^{-.25(x^2 + y^2)}$
5. $\frac{1}{x} + \frac{1}{y} + xy$

Locating Extrema for Functions with Constraints

Most optimization problems do not require an answer for all values of the variables, but are looking for an optimum value subject to a particular constraint. Consider the following problem.

Find a triangle such that the product of the sines of its angles is maximal.

This problem can be reformulated as

Maximize: $f(x, y, z) = \sin(x) \sin(y) \sin(z)$

Subject to: $x + y + z = \pi$

The angles are constrained to be angles in the same triangle. Thus, the constraint on the sum.

We will solve this problem as a staged decision problem[9] by looking at the constraint on the sum of the angles as a resource to be processed, and x, y, and z as decision variables. What we will do is look at the problem as one of making a decision on how to allocate the resource of π radians among the three angles of the triangle. The decision will be made in three stages, namely x, y, and z. At each stage we build up the product of the sines. The process can be diagrammed as follows

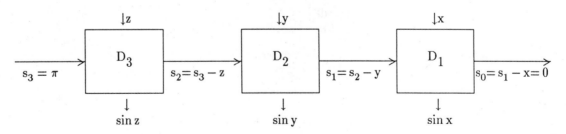

[9]For more details on this method of problem solving, see G. L. Nemhauser, *Introduction to Dynamic Programming*, John Wiley and Sons, New York, N.Y.: 1966.

We can think of the resource (radians to be allocated to the angles) as flowing from left to right through three decision processes. Each decision process has an input, the size of the angle, and an output, the sine of the angle. Also, each decision depletes the resource. At each stage of the process we have only one variable to worry about, and the last stage is really no decision at all. We must allocate whatever is left of the resource to x.

Thus, we have broken the process into three linked maximization problems. We will number the problems according to the numbers of the decisions in our diagram and subscript the results with these numbers. The three problems are given below.

$$f_1(s_1) = \max_{x = s_1} \{\sin x\}$$

$$f_2(s_2) = \max_{0 \le y \le s_2} \{\sin y \cdot f_1(s_1)\} = \max_{0 \le y \le s_2} \{\sin y \cdot f_1(s_2 - y)\}$$

$$f_3(s_3) = \max_{0 \le z \le s_3} \{\sin z \cdot f_2(s_2)\} = \max_{0 \le z \le s_3} \{\sin z \cdot f_2(s_3 - z)\}$$

We will start with the first problem and solve these three problems in sequence. We begin with the first. Since we have no choice in the first problem,

$$f_1(s_1) = \sin(s_1)$$

Given this result, we move to the second problem

$$\max_{0 \le y \le s_2} \{\sin y \cdot \sin(s_2 - y)\}$$

We solve this problem in the traditional way. Authoring the expression $\sin(y)\sin(S-y)$ and differentiating, we see that

$$\frac{df_2}{dy} = \cos(y)\sin(s_2 - y) - \sin(y)\cos(s_2 - y)$$

or, using elementary identities,

$$\frac{df_2}{dy} = \sin(s_2 - 2y) .$$

Setting this equal to zero and checking the second derivative, we see that for our conditions f_2 attains a maximum value for $y = \frac{s_2}{2}$. Thus,

$$f_2(s_2) = \sin^2\left(\frac{s_2}{2}\right)$$

We are now ready to solve the last of our three problems:

$$\max_{0 \le z \le \pi} \{\sin z \cdot \sin^2\left(\frac{\pi - z}{2}\right)\}$$

Note that we substituted π for s_3 in the above problem. This is for convenience when using DERIVE. After entering the function in brackets and differentiating with respect to z, we obtain a rather complicated expression involving sines and cosines. An attempt to solve this equation in Exact mode does not yield an answer. If we graph the expression for the derivative, we see that it does have several zeros. In fact, there is one fairly close to 1. Returning to the algebra window and using the Manage/Precision option, set the precision to Approximate and attempt to solve the equation again. Set the bounds on the answer to 0 and $\pi/2$. This yields a result of 1.04719.

In order to make some sense of this number, divide it by π. The result is .333333, or what appears to be $\frac{1}{3}$. This would mean that the optimal value is attained when $z = \frac{\pi}{3}$. We can check this by returning to Exact mode and substituting this value for z into the expression for the derivative. The result after simplifying is 0. If we take the second derivative and substitute $z = \frac{\pi}{3}$, the result is a negative number. This confirms our hypothesis that a maximum value for the product of the sines of the angles is attained when $z = \frac{\pi}{3}$.

In order to complete the analysis, note that $s_2 = \pi - z = \frac{2\pi}{3}$ and that $x = y = \frac{s_2}{2} = \frac{\pi}{3}$. This means that the maximum product of the sines of the angles occurs when

$$x \;=\; y \;=\; z \;=\; \tfrac{\pi}{3}$$

or when we have an equilateral triangle.

Note that at each stage of the solution of this problem, we had a relatively simple single variable optimization problem to solve. That is the beauty of looking at the problem as a staged decision process and starting with the last decision.

The technique of solving a multivariable optimization problem with a single constraint, looking at it as a staged decision problem, and solving it by starting with the last stage of the problem and working back to the original problem is called *dynamic programming*. It was invented by the American mathematician Richard Bellman, and has many interesting applications. A student interested in further applications of this technique is encouraged to explore books on operations research.

The Laboratory Report

Show how the following constrained minimization problem can be viewed as a staged decision problem.

$$\text{minimize } x^2 + y^2 + z^2$$

$$\text{subject to: } x + y + z \geq 1 \ ; x, y, z \geq 0$$

Solve this problem as a sequence of linked minimization problems.

Work Sheet

Laboratory #20

1. Fill in the process diagram for the following decision problem:

$$\text{minimize } x^2 + y^2 + z^2$$

$$\text{subject to: } x + y + z \geq 1 \ ; \ x, y, z \geq 0$$

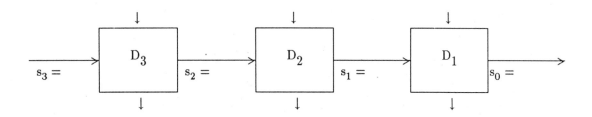

2. Identify and solve each of the resulting linked minimization problems.

3. Change the constraint to $x + y + z = 1$ and use this to solve for z in terms of x and y. Substitute this expression for z into the objective function $x^2 + y^2 + z^2$ and do a surface plot of the result. Locate a minimum point on this surface. Compare your observation to the answer you obtained in exercise 2 above.